RAND NATIONAL DEFENSE RESEARCH INSTITUTE

T0210384

# National Guard Youth ChalleNGe

## Program Progress in 2015–2016

Jennie W. Wenger, Louay Constant, Linda Cottrell, Thomas E. Trail, Michael J. D. Vermeer, Stephani L. Wrabel

Prepared for the Office of the Secretary of Defense

For more information on this publication, visit www.rand.org/t/RR1848

**Library of Congress Cataloging-in-Publication Data** is available for this publication.
ISBN: 978-0-8330-9834-4

Published by the RAND Corporation, Santa Monica, Calif.
© Copyright 2017 RAND Corporation
**RAND**® is a registered trademark.

*Cover: Jim Greenhill for the United States Government.*

Support RAND
Make a tax-deductible charitable contribution at
www.rand.org/giving/contribute

www.rand.org

# Preface

The National Guard Youth ChalleNGe program is a residential, quasi-military program for youth between the ages of 16 and 18 who are experiencing difficulty in traditional high school. The program is operated by participating states through their state National Guard organizations with supporting federal funds and oversight. The first ChalleNGe sites began in the mid-1990s; today there are 40 ChalleNGe sites in 29 states, the District of Columbia, and Puerto Rico. To date, more than 145,000 young people have completed the ChalleNGe program. Congress requires the ChalleNGe program to deliver a report on its progress each year.

The program includes a 5.5-month Residential Phase followed by a 12-month Post-Residential Phase, with support from a mentor. The stated goal of ChalleNGe is "to intervene in and reclaim the lives of 16–18 year old high school dropouts, producing program graduates with the values, life skills, education, and self-discipline necessary to succeed as productive citizens."

In this report, we provide information in support of the required annual report to Congress. We also lay out a framework for use in evaluating ChalleNGe sites; subsequent reports will provide additional information on future cohorts of students and will build on this framework to develop more detailed and more effective metrics, and will provide strategies for data collection in support of these metrics. Methods used in this study include site visits, data collection and analysis, literature review, and development of two tools to assist in improving the metrics—a theory of change (TOC) and a program logic model.

This report will be of interest to ChalleNGe program staff, to personnel providing oversight for the program, and to policymakers concerned with designing effective youth programs and/or determining appropriate metrics by which to track progress in youth programs.

This research was sponsored by the Office of the Assistant Secretary of Defense for Manpower and Reserve Affairs and conducted within the Forces and Resources Policy Center of the RAND National Defense Research Institute, a federally funded research and development center sponsored by the Office of the Secretary of Defense, the Joint Staff, the Unified Combatant Commands, the Navy, the Marine Corps, the defense agencies, and the defense Intelligence Community. For more information on the RAND Forces and Resources Policy Center, see http://www.rand.org/nsrd/ndri/centers/frp.html or contact the director (contact information is provided on the web page).

Comments or questions on this draft report should be addressed to the project leader, Jennie Wenger, at Jennie_Wenger@rand.org.

# Contents

# Figures

# Tables

# Summary

The National Guard Youth ChalleNGe program is a residential, quasi-military program for youth between the ages of 16 and 18 who are experiencing difficulty in traditional high school. The program is operated by participating states with supporting federal funds and oversight through their state National Guard organizations. The program began in the mid-1990s; today there are 40 sites in 29 states, the District of Columbia, and Puerto Rico. To date, more than 190,000 young people have taken part in ChalleNGe, with more than 145,000 completing the program.

ChalleNGe's mission is to intervene in and reclaim the lives of 16–18-year-old high school dropouts, producing program graduates with the values, life skills, education, and self-discipline necessary to succeed as productive citizens. ChalleNGe delivers a congressionally mandated report on the program's progress each year; past reports have included information on total and average spending by program and metrics of cadets' scores on standardized tests, as well as details on the number of participants and some information on participants' post-ChalleNGe placement (postsecondary school, civilian labor market, military service, etc.).

The ChalleNGe program emphasizes eight core components, many of which involve developing noncognitive or socioemotional skills. Previous research has found that ChalleNGe has a positive influence on participants' long-run labor market outcomes and is cost-effective (Millenky et al., 2011, Perez-Arce et al., 2012). However, there has been substantial variation across sites, even after controlling for basic metrics of participants' backgrounds. This variation extends to a fairly broad list of characteristics. For example, credentials awarded vary across the sites, and the timing of cycles is inconsistent (many classes begin in January and July but others begin at other times). Further, research has not addressed longer-term outcomes of the program. Attaining a better understanding of the source(s) of variation and understanding a fuller range of outcomes would assist program staff in determining how best to utilize resources and better inform stakeholders on how well ChalleNGe is achieving its mission.

The purpose of the RAND Corporation's project is twofold. We focused initially on gathering and analyzing existing data in support of the 2016 annual report to Congress. Then, to support future analyses, our team will focus on developing a rich and detailed set of metrics to capture more information about the differences between individual sites; data using these metrics will contribute to future annual reports and will allow us to explore variation across sites in more detail. Methods used in this study include site visits, data collection and analysis, literature review, and development of two logic models. This report, the first in a series, provides a snapshot of ChalleNGe in 2015–2016 and reports on the initial steps in the process of developing a richer set of metrics to measure long-term outcomes.

## Cross-Site Metrics for the 2015 ChalleNGe Class

Our data collection describing the ChalleNGe classes of 2015 provides a snapshot of recent performance. Most of the metrics collected by ChalleNGe sites to date focus on inputs, activities, and outputs, with a few metrics of shorter-term outcomes.

We placed considerable focus on one existing metric, the Tests of Adult Basic Education (TABE), which serves as the primary metric of academic progress among ChalleNGe participants. Although the test is generally appropriate for this purpose, the average grade equivalent scores reported by ChalleNGe sites do not indicate the number or proportion of cadets who have reached key benchmarks. Fortunately, TABE-based benchmarks exist. We presented two metrics linked to ChalleNGe-relevant outcomes: achieving a level of at least grade 9 (early high school) and achieving a grade level of at least grade 11 (late high school). The data indicate that cadets make considerable progress in a number of areas while attending ChalleNGe. We found that cadets who enter the program scoring at the middle school level or above are quite likely to achieve key benchmarks by graduation. If combined with a metric based on test score growth, reporting benchmarks achieved could provide a much more complete picture of ChalleNGe cadet performance, with little, if any, additional information collection required.

Within the ChalleNGe program, placement is considered a key metric. ChalleNGe staff work to keep in contact with graduates and their mentors, both to assist the graduates in finding opportunities and to record the graduates' activities. Placements may include military service, additional education, or working (as well as combinations of these, such as attending school and working). Half of graduates who report having a placement six months after graduation are obtaining additional education; many of the rest are employed, with smaller numbers serving in the military or reporting a combination of placements or some other sort of placement. Overall, the ChalleNGe placement rate of 72 percent resembles the activities of high school diploma graduates and exceeds the placement rate of General Educational Development (GED) holders in the High School Longitudinal Study of 2009 (HSLS:09).

Our analysis of cost data provided by the sites indicates that while most ChalleNGe sites have somewhat similar average costs per graduate, a few sites have costs that are much higher. We explored several possible reasons for cost variation and found that, rather than differences in sites' ages or credentials awarded, size (number of graduates) is the driver of cost. Of course, we would expect costs to vary with the number of cadets, but sites that have fewer than 150 graduates per year have substantially higher costs than larger sites, while costs per graduate generally are quite similar at sites that have at least 150 graduates. This suggests that the fixed costs of running a ChalleNGe program dominate other costs in smaller sites. While these small sites are responsible for only about 6 percent of total costs, the data indicate that encouraging sites to attain a size of at least 150 graduates has the potential to improve cost-effectiveness.

## An Initial Framework for Measuring Long-Term Outcomes

To begin the process of improving program metrics and measuring longer-term impacts, we developed two tools: a theory of change (TOC), which describes the mechanisms underlying ChalleNGe; and a program logic model, which describes the relationships between resources, activities, and outcomes.

The TOC for the ChalleNGe program posits that an intensive, residential-based, regimented program, scaffolded by mentorship after program completion, will increase the likelihood that at-risk youth can turn around their lives and achieve success in work and life. The TOC also provides a foundation for identifying the types of outcomes that will be measured to track progress. The TOC includes five tenets that contribute toward helping a young person achieve a rewarding, productive life, which are based on the eight core components of the program:

1. Develop leadership and followership behaviors through discipline, hard work, and persistence.
2. Engage in activities that promote good physical health.
3. Act as a responsible citizen and build strong linkages to the community through service and participation.
4. Attain academic skills and credentials to create job-readiness and the potential for success in the labor market.
5. Strengthen socioemotional skills to build life-coping strategies.

The program logic model specifies the reasoning behind program structure and activities and how those activities are connected to expected program results. Program inputs (the resources needed to administer the program) include policy and planning materials to guide program activities and identify the assets needed to house and instruct cadets. Program activities include Acclimation Period orientation activities undertaken to prepare cadets for ChalleNGe (e.g., performing physical exams as well as instructing cadets on program standards and expectations). The Acclimation Period activities feed directly into program activities during the Residential Phase. Program outputs include those related to cadet instruction activities (e.g., housing, instructing, and mentoring cadets) and those related to the end process of graduating cadets (e.g., administering standardized tests, awarding credentials, and placing cadets). Outcomes expected to result from program completion include those in the short term (within three years of graduation), medium term (within three to seven years of graduation), and long term (seven years or more after graduation). These include positive outcomes for the cadets themselves and their families (e.g., better job skills and job prospects), as well as for their communities, government, and the military (e.g., an increase in individuals participating in community service activities, greater tax revenue, and increased military enlistment from underrepresented populations). Understanding the dynamic flow of the relationships between and among the inputs, outputs, and outcomes, and measuring the expected connections among these components will allow for systematic evaluations of the ChalleNGe program (W. K. Kellogg Foundation, 2006).

## Data Collection: Barriers and Strategies

The tools just described are useful for understanding the types of metrics and data collection efforts necessary to measure the longer-term impacts of the program and for communicating program goals to stakeholders. In general, these models indicate that effectively linking aspects of the ChalleNGe program to longer-term outcomes will likely require additional data collection efforts. Adult education programs collect some relevant longer-term outcomes on their participants and some of their data collection strategies are relevant here. For example, in many cases, adult education programs utilize existing administrative datasets.

While such datasets contain information relevant to ChalleNGe, the ChalleNGe program faces several barriers to such data collection strategies. Barriers include mobility of participants, the large number of sites in multiple states, and the lack of formal linkages between ChalleNGe sites and relevant state departments. For these reasons, leveraging administrative datasets represents a costly strategy in terms of establishing official data use agreements. Fortunately, the ChalleNGe sites have counselors in place to collect some data, which could be fine-tuned to represent better metrics. Surveys of past cadets also appear to represent a viable method of collecting additional information.

## Conclusion

In summary, the ChalleNGe model appears well grounded in the existing literature, and the data that we have collected for this report indicate that cadets across ChalleNGe sites made substantial progress in multiple areas. However, most of the metrics collected so far do not include information necessary to measure the longer-term outcomes and impacts of the program. Future reports will focus on both developing new metrics and discerning relevant time trends as we continue to collect data across the ChalleNGe sites.

# Acknowledgments

We are grateful to the administrative staff of the National Guard Youth ChalleNGe program who responded to our data request in a timely fashion and provided substantial amounts of background and contextual information on ChalleNGe sites throughout the course of planning and completing this report.

We are also grateful to our RAND colleagues for their support: John Winkler, Craig Bond, and Lisa Harrington reviewed various portions of the report; Kristin Leuschner, Barbara Bicksler, Arwen Bicknell, and Beth Bernstein worked to improve the appearance and clarity of the report; Neil DeWeese, John Tuten, and Cynthia Christopher provided administrative support. Gabriella Gonzalez of RAND, David DuBois of the University of Illinois at Chicago, and Michael MacLaren provided reviews that ensured our work met RAND's high standards for quality.

We thank all who contributed to this research or assisted with this report, but we retain full responsibility for the accuracy, objectivity, and analytical integrity of the work presented here.

# CHAPTER ONE

# Introduction: The National Guard Youth ChalleNGe Program

The National Guard Youth ChalleNGe program is a residential, quasi-military program for young people between the ages of 16 and 18 who have left high school without a diploma or are at risk of dropping out. ChalleNGe's mission is "to intervene in and reclaim the lives of 16–18 year old high school dropouts, producing program graduates with the values, life skills, education, and self-discipline necessary to succeed as productive citizens."[1] The program's vision is to be recognized as America's premier voluntary program for 16–18-year-old high school dropouts, serving all U.S. states and territories.

ChalleNGe is based on eight core components (leadership/followership, responsible citizenship, service to community, life-coping skills, physical fitness, health and hygiene, job skills, and academic excellence). The program is operated by participating states through their state National Guard organizations with supporting federal funds and oversight. The National Guard is responsible for all the day-to-day operational aspects of the program; the Office of the Secretary of Defense provides oversight. States are required to contribute at least 25 percent of the operating funds. The first ten ChalleNGe sites began in the mid-1990s; today there are 40 ChalleNGe sites in 29 states, the District of Columbia, and Puerto Rico, with more than 145,000 young people having completed the ChalleNGe program to date. Appendix A includes a complete list of ChalleNGe sites. Several programs have opened recently or are still in the process of opening. In this report, we focus on the 37 programs that have been operational long enough to provide data on graduates; future reports will also include information on the newest programs.

ChalleNGe delivers a congressionally mandated report on the program's progress each year; past reports have included information on total and average spending across the program and metrics of cadets' scores on standardized tests, as well as details on the number of participants and some information on participants' postresidential placement (postsecondary school, civilian labor market, military service, etc.).

Previous research has found that ChalleNGe has a positive influence on participants' long-run labor market outcomes and is cost-effective (Millenky et al., 2011, Perez-Arce et al., 2012). However, there has been substantial variation in outcome metrics across sites, even after controlling for basic metrics of participants' backgrounds. Such variation in outcomes might be due to variation in individual sites. While the overall ChalleNGe program is structured, individual sites have substantial discretion in deciding how to carry out the components. In

---

[1] See, for example, the ChalleNGe website (National Guard Youth ChalleNGe, n.d.). The mission statement appears to be widely shared across ChalleNGe sites. It is quoted in various materials and briefings used at the sites and was included in briefings that formed part of our site visits.

particular, there is variation in the academic component; some sites focus on preparing cadets to take the General Educational Development (GED) exam, others award high school credits, and some award high school diplomas. Different sites also have developed unique models for focusing on the nonacademic components of the program. Further, research has not addressed longer-term outcomes of the program, such as postsecondary degree attainment. Attaining a better understanding of the source(s) of variation and understanding a fuller range of outcomes would assist program staff in determining how best to utilize resources and better inform how well ChalleNGe is achieving its mission.

The purpose of the RAND Corporation's project is twofold. We focused initially on gathering and analyzing existing data in support of the 2016 annual report to Congress; findings from this analysis are contained in this report. Then, to support future analyses, our team will focus on developing a rich and detailed set of metrics to capture more information about the differences between individual sites. We will collect data and provide analyses on these new metrics; this information will contribute to the program's annual reports in 2017, 2018, and 2019. An initial framework for these future analyses is contained in this report.

In the remainder of this chapter, we provide additional background and review existing research on the ChalleNGe program. We then describe in more detail the focus of this report and the methodology we used. We conclude with a roadmap for the remainder of the report.

## The ChalleNGe Model

The ChalleNGe program has several unique characteristics. Participants (referred to as *cadets*) attend a site located in the state where they live. Participation is voluntary and there is no tuition cost to the cadet or his or her family, although cadets must apply to the program and most sites require a "packing list" of items to be purchased by the cadet/family and brought to the program site on the first day of the program. There is some variance among sites in the application process, but it generally involves an applicant filling out an application, taking a standardized test (the Tests of Adult Basic Education [TABE]), and completing an interview (or attending an information session). Most sites do not have test score requirements. Additionally, applicants must not be awaiting sentencing, on parole or on probation for anything other than a juvenile offense, and they must not be under indictment, accused, or convicted of a felony.[2]

While taking part in the initial 5.5-month portion of ChalleNGe, cadets reside at the site. During this time, cadets wear uniforms, live in a barrack-like atmosphere, and perform activities generally associated with military training (e.g., marching, drills, physical training). The first two-week phase of the program, referred to as the Acclimation Period, is designed to allow new cadets time to adjust to the new environment and the expectations that the ChalleNGe program requires for success; coursework begins at the end of the Acclimation Period. For the next five months, cadets attend classes during much of the day; sites may focus on the completion of a GED or High School Equivalency Test (HiSET) credential. Cadets also have the option to earn high school credits that they can use to transfer to a high school at the end of ChalleNGe and subsequently go on to earn a high school diploma. Depending on the program, cadets can earn some combination of the above. (Some ChalleNGe sites even award state-certified high school diplomas.) Not all cadets complete the residential portion

---

[2]  DoD Instruction 1025.8, March 20, 2002.

of the ChalleNGe program (successful completion is referred to as *graduation*). Most cadets who leave ChalleNGe prior to graduation choose to withdraw, but sites can and do dismiss cadets who violate key policies. Cadets are not enlisted in the military while participating in the Residential Phase of ChalleNGe, and there is no requirement of military service following completion of the program.

ChalleNGe places considerable focus on the development of noncognitive or socioemotional skills such as leadership/followership, having positive interpersonal relationships, developing goals and detailed plans to accomplish the goals, anger management, and attention to detail, among others. Indeed, the basis of the program is the following eight core components:

- leadership/followership
- service to community
- job skills
- academic excellence
- responsible citizenship
- life-coping skills
- health and hygiene
- physical fitness.

Each ChalleNGe site is charged with developing cadets' skills and abilities in all of these areas. Mentorship plays a key role—each cadet has a mentor, and the relationships between cadet and mentor continue for at least 12 months after the cadet graduates from the Residential Phase of ChalleNGe (through the Post-Residential Phase). Somewhat unique among mentoring programs, the ChalleNGe mentoring model is youth-initiated; cadets are required to nominate mentors. Mentors, who receive in-person training from ChalleNGe staff, are volunteers. Mentors are encouraged to maintain regular contact with ChalleNGe cadets during the program and for at least one year after the cadet completes the program; mentors also maintain contact with program staff throughout the Post-Residential Phase.

## Previous Research on ChalleNGe

There has been previous research to evaluate the effects of ChalleNGe, although studies to date have not accounted for variation in outcomes across different program sites or examined some longer-term outcomes associated with the program.

The relationship between education and eventual labor market success is well established and robust.[3] Based on a random control trial including a limited number of ChalleNGe sites, the ChalleNGe program has been shown to have positive impacts on labor market outcomes for those who participate.[4] When compared with similar young people not in the program,

---

[3]  For a detailed review, see Card (1999).

[4]  A random control trial, considered the gold standard within social science research, compares outcomes for two groups: a *treatment* group (in this case, a group of applicants who were accepted into the ChalleNGe program) with a *control* group (a similar group of young people who were not admitted into the ChalleNGe program). Randomization between the treatment and control groups is a key requirement for a random control trial; with randomization, differences between the groups can be attributed to the ChalleNGe program. Note that in the case of the ChalleNGe program, the random control trial included a subset of program sites.

ChalleNGe cadets at these sites completed more postsecondary education and were more likely to be working three years after entering the program, but it is worth noting that no effects were found on a number of outcomes that might be expected to respond to the ChalleNGe model (e.g., arrest rates), and those who entered ChalleNGe had more negative outcomes in a few cases such as overweight status (Millenky et al., 2011).

However, there is evidence that the life-coping skills stressed in ChalleNGe appear to increase participants' noncognitive or socioemotional skills (Malone and Atkin, 2016). The relationships among many other aspects of the core components and long-term outcomes are less well established, but there is evidence of the effectiveness of mentoring, especially when the mentoring relationship is structured as it is within ChalleNGe.[5]

The labor market outcomes are impressive when compared with studies of other programs aimed at high school dropouts, and are perhaps even more impressive due to inadvertent timing. The random control trial of ChalleNGe described by Millenky et al. (2011) took place during the recent severe economic recession whose effects were particularly pronounced on young workers with minimal amounts of education. Indeed, a separate and careful analysis of all costs and benefits based on the outcomes from the random control trial found that ChalleNGe is cost-effective, producing approximately $2.66 in benefits (appropriately discounted) for each $1.00 invested (Perez-Arce et al., 2012). This cost-benefit analysis constitutes positive findings for the program and includes longer-term outcomes than most sites collect (three years after graduation). But it is noteworthy that this research includes only a subset of ChalleNGe sites.

Despite these positive outcomes for the program, outcome metrics differ across sites, and there are indications that some program attributes vary across sites as well. For example, graduation rates differ across the sites, even after controlling for basic metrics of participants' backgrounds and initial preparation (Wenger et al., 2008). Some of this variation may be related to differences in state requirements or education systems; other variation could be driven by differences in populations, population densities, and local labor markets of the states and areas where the sites are located.

In summary, past research has found that participants in the ChalleNGe program have more favorable labor market outcomes in the immediate years after entering ChalleNGe, but this research has examined a relatively small number of longer-term outcomes, and only for cadets who attended a subset of programs. There is some evidence that the program influences other skills, such as life-coping skills, and the mentorship component of the program appears to have a positive impact. However, there is substantial variation across the sites on a number of characteristics; for example, programs vary in terms of credentials awarded, size, and of course state attributes. Moreover, graduation rates differ substantially across programs. Attaining a better understanding of the sources of these differences would assist policymakers in formulat-

---

[5] The ChalleNGe youth-initiated mentoring model is linked to more enduring relationships between the cadet and the mentor, especially among cadets who select the mentors themselves. Schwartz et al. (2013), based on 21- and 38-month follow-up surveys at one ChalleNGe program, found that higher rates of enduring cadet-mentor relationships were in place at the 21-month follow-up in cases where the cadet selected the mentor compared to those cases where the parents or ChalleNGe staff selected the mentors. Notably, overall rates of contact were lower among all selection approaches at the 38-month follow-up. The study did find that cadets who reported remaining in touch with their mentor at the 38-month follow-up were more likely to have attained a GED or high school diploma, achieved college credit, worked and reported higher earnings, and had fewer criminal convictions. These effects were not as strong among cadets who reported remaining in touch with a mentor at the 21-month but not 38-month follow-up, and no different from zero for cadets who did not report a mentor relationship in the 21-month follow-up compared to the control group. For more information on the effects of mentoring across various interventions, see Rhodes et al. (2006) and Tierney, Grossman, and Resch (2000).

ing policies and allocating resources to achieve greater impact, and could also assist decision-makers within the ChalleNGe program and at individual sites as they work to improve the design and implementation of the program. Collecting consistent data over several years will allow us to explore the effects of these differences.

## Focus of This Report

This report, the first in a series for our project, serves two purposes. The first is to provide a snapshot of the ChalleNGe program during 2015–2016.[6] The second is to begin the process of developing a richer set of metrics that will describe the long-term effects ChalleNGe has on participants after they leave the program, and thus will serve to measure the extent to which the ChalleNGe program is achieving its mission.

To support the first purpose, we include information gathered from individual ChalleNGe sites detailing the number of participants who began and completed ChalleNGe, as well as metrics of their academic progress, improvements in physical fitness, and service to community. This information meets the program's current annual reporting requirements and will be used in the program's 2016 report to Congress.[7] Similar information was collected for past reports, which included metrics of average gains on standardized test scores, some information on community service and physical fitness gains, and the number of graduates who are *placed* (enrolled in additional education, employed, or serving in the military).[8] However, those reports did not include an indication of the number of cadets who achieved a given academic level; here we develop more detailed metrics of academic progress based on existing data. Past reports also did not include information on additional credits or degrees attained, earnings, or job stability of graduates. While data on such metrics are not currently collected, these metrics are relevant to participant success as defined in the program's mission statement.

To address the second purpose, we lay out an initial framework for measuring the longer-term outcomes of the program (such as additional credits or degrees attained, earnings, or job stability of graduates). Because the program's mission focuses on participants' success as adults (after completion of the program), many of these metrics will focus on longer-term outcomes. Developing metrics that are linked to longer-run outcomes will make it possible to determine the extent to which the program is achieving its mission.[9] In this report, we undertake the initial step in developing metrics by laying out a framework describing how the ChalleNGe program works. We have developed a theory of change (TOC) to conceptually describe the ChalleNGe program and its expected impact. The TOC informs the logic model, an operational tool to guide the development of metrics to monitor progress toward achieving the program's central goals and evaluate its effectiveness. The logic model will begin to spell out the process through which change will occur, the intermediate steps involved, and the long-term outcomes (Anderson, 2005). We also discuss barriers and strategies for data collection to support improved measurement.

---

[6]  Because different sites run on different schedules, this report includes some information on sites that began in 2014, as well as some information on a few sites completed in 2016; see Figure A.1 in Appendix A for program schedules.

[7]  See 32 U.S.C. §509(k) for annual reporting requirements.

[8]  See, for example, the 2015 annual report (National Guard Youth ChalleNGe, 2015).

[9]  In some ways, the ChalleNGe program faces particular hurdles in terms of data collection. We discuss these, and strategies to overcome them, in Chapter Four.

## Methodology

Given the multiple goals of this report, we combine several methodologies. To meet the first purpose—documenting progress and supporting the annual report to Congress—we collected information from each ChalleNGe site. This program-level information is typical of what has been included in past annual reports. We reviewed information from each site on program characteristics; 2015 budget and sources of funds; number of applicants, participants, and graduates; credentials awarded; and metrics of physical fitness and community service/engagement. We also reviewed information on staffing, dates classes began and ended, and postresidential placements. We requested and received the information through secure data transfer (although we requested no identifying information). We specified that sites should include information from the two classes that began during 2015.

In this data collection, we also requested cadet-level information on graduation, credentials awarded, changes in TABE grade equivalent scores, as well as placements during the Post-Residential Phase.[10] Past reports included only site-level metrics, such as the average gain in TABE grade equivalent scores or the number of cadets placed. *Average gain* in TABE grade equivalent scores is widely used but problematic (Lindholm-Leary and Hargett, 2006);[11] achieving *key levels* on the TABE predicts other relevant outcomes, such as passing the GED exam. We used the cadet-level information to develop and report metrics based on achieving key TABE levels.

To meet the second purpose of this report—beginning to amass a richer set of metrics that are tied to longer-term outcomes and the program's mission—we first developed two tools: a TOC and a program logic model. We developed these tools or models based on information gleaned from two site visits to ChalleNGe sites (the Mountaineer ChalleNGe Academy in West Virginia and the Gillis Long ChalleNGe site in Louisiana) and from other program documents. We chose these programs based partly on convenience, but also based on differences in size, region, and academic focus at these sites. The two site visits included meetings between two or three RAND researchers and program staff, including the site director, commandant of cadets, recruiting coordinator, counseling staff member, residential activities coordinator, academic instruction leader, postresidential mentor coordinator, and a member of the cadre.[12] The meetings covered the programs' outreach, application, and selection process; cadet Acclimation Period orientation activities; Residential Phase activities; classroom/academic activities; postresidential activities and mentoring; and the overall goals of the programs and potential metrics to assess progress toward meeting those goals. We also used information gathered from a 2016 meeting of some ChalleNGe directors in which they posed ideas for program metrics, as well as information from the relevant literature.[13] Measuring longer-term outcomes will require collecting different data, in particular more data on cadets who have completed ChalleNGe.

---

[10] TABE scores measure academic achievement in math and language arts and constitute a key metric for the ChalleNGe program. They are reported in past analyses; see, for example, the 2015 annual report (National Guard Youth ChalleNGe, 2015).

[11] The problematic nature of grade equivalent scores for measuring changes is explained in Chapter Two.

[12] *Cadre* (collective noun) is the term used to describe staff members who are in charge of groups of cadets at all times. Cadre accompany cadets to meals, physical training, and generally even to class; some cadre are on duty at night as well.

[13] Combining information from such sources is a typical methodology for building or developing logic models; see Knowlton and Phillips (2009).

With this in mind, we identified some data strategies and barriers that are especially relevant to the ChalleNGe program. Future reports will focus on developing specific outcomes that tie to the program's mission and on collecting information to measure these longer-term outcomes.

## Organization of This Report

The remainder of the report consists of three chapters:

- Chapter Two provides a snapshot of the ChalleNGe program in 2015–2016 and supports the annual report to Congress. It includes information from recent classes that is comparable to what was included in past reports, as well as information on the proportion of cadets meeting key TABE levels, placement rates, and analyses on program costs.
- Chapter Three discusses our initial framework for measuring the longer-term outcomes of the program. This chapter presents the TOC and program logic model and outlines recommendations and plans for future metrics and data collection efforts.
- Chapter Four closes with concluding thoughts.

# CHAPTER TWO

# Data and Analyses: 2015 ChalleNGe Classes

The analyses in this chapter are based on information collected from the individual ChalleNGe sites in the fall of 2016. We requested information on classes that began in 2015 (generally referred to as Class 44 and Class 45 across the sites). Some sites began operating quite recently and therefore lack information on past classes. In particular, North Carolina's New London site and the Texas-East ChalleNGe site began operating in the middle of 2015; Class 45 was the first class at each of these sites. Therefore, they had no cadets during Class 44 and thus reported no Class 44 data.

We begin by discussing the TABE, which is used by all sites and is a central metric of academic success at ChalleNGe. We then present statistics for the program from 1993 to 2015, and describe cross-site metrics for the 2015 class.

We next discuss our cross-program analyses of TABE, placement, and costs. For consistency with earlier reports, we collect and present information on TABE score gains, but also present information based on our improved TABE metrics. Finally, we discuss how the program compares to similar programs. Appendix C provides detailed information organized by ChalleNGe site.

## Tests of Adult Basic Education

ChalleNGe cadets take the TABE at the beginning of the program and again at the end of the Residential Phase of the program.[1] The TABE is one of the three most commonly used assessments in adult basic and secondary education (U.S. Department of Education [USDOE], Office of Career, Technical, and Adult Education [OCTAE], 2015). The TABE offers three types of scoring information: a number of correct responses, a scale score, and a grade equivalent score.[2] ChalleNGe sites traditionally have reported TABE scores in terms of grade

---

[1]  Some sites use the TABE more extensively to track progress over the course of the five-month Residential Phase of ChalleNGe. The TABE is designed both for formative (placement) and summative (progress or gains) assessment. For more details on the TABE, see Appendix B.

[2]  Receiving a grade equivalent score of 5.9 indicates that in current test administration the student's performance was similar to that of an individual performing at the 50th percentile of students who were in the ninth month of fifth grade, while a score of 7.1 suggests the individual is performing similar to students at the 50th percentile in the first month of the seventh grade.

Measurement errors associated with using the TABE may impact the measures of cadets' gains over the course of the program. The TABE comes with recommended protocols and specific tools (e.g., a "Locator" assessment) to ensure that participants are given the appropriate tests and that the findings from these tests are accurate. Deviation from the intended

equivalents. Indeed, average TABE gain scores (change in average grade equivalent scores over the course of the program) have been featured prominently in past annual reports.

Past ChalleNGe data indicate that cadets typically gain two or more years on the grade equivalent score over the course of the Residential Phase. Although such gains certainly imply substantial academic progress, there are several drawbacks to this metric. First, measuring gains in (i.e., subtracting) grade equivalent scores and averaging them is inappropriate. Given the way in which grade equivalents are calculated, the gain metric inaccurately identifies the amount of growth or change experienced (Lindholm-Leary and Hargett, 2006; Jacob and Rothstein, 2016). We recognize that averaging grade equivalent scores and reporting the gains are common practices in the Adult Basic Education/Adult Secondary Education (ABE/ASE) world. But even if one accepts the measurement issues associated with grade equivalent scores, average grade equivalent gains do not reveal the extent to which all, most, or only a few cadets make substantial progress over the course of the program. Second, the TABE has a "ceiling"— the highest possible score is a 12.9. Therefore, for a cadet who scores close to 12.9 on the initial test, the TABE can demonstrate only a limited amount of progress. Finally, there is no research linking or mapping the TABE gains to other outcomes of interest.

Fortunately, there are ways of addressing these issues and providing more relevant information using currently available data.[3] Grade equivalent and scale scores *can* be linked to some outcomes of interest. TABE scores have been linked to the Scholastic Aptitude Test (SAT), the Enhanced ACT (Standardized College Entrance Exam, formerly known as American College Testing), and other college placement tests. For example, SAT scores well below the mean would be expected by students whose scores indicated they had not yet reached the end of twelfth grade on the TABE (West Virginia Department of Education, n.d.).[4] Scale scores on the TABE have been linked to scale scores on the GED (Olsen, 2009), and grade equivalent scores have been linked to Armed Forces Qualification Test (AFQT) scores (Wenger, McHugh, and Houck, 2006).

Finally, linkages can be made between the TABE and the likelihood of passing the GED using the National Reporting Service for Adult Education (NRS)'s six Educational Functioning Levels and associated scores from the Comprehensive Adult Student Assessment System (CASAS) examination (NRS, 2015; CASAS, 2003). What these linkages suggest is that reaching a 9.0 grade equivalent score on the TABE is associated with a pass rate of 70 percent or higher on the Reading, Language Arts, and Math sections of the GED test. For individuals reaching an 11.0 grade equivalent score on the TABE, that pass rate increases to 85 percent or higher.[5] More recent work from CASAS (2016) further confirms that individuals with a 9.0

---

TABE protocols may form an additional source of measurement error on cadet performance. More information on the TABE is presented in Appendix B.

[3]  Potential approaches to addressing the current TABE measurement issues include taking the average entry scale score and the average exit scale score and converting those values to grade equivalents; or averaging scale scores and comparing average change in scale scores across the program locations. Finally, it is possible to compute *reliable change* scores that indicate how many test-takers have made improvement after allowing for measurement error; such scores may be especially helpful in discerning gains of those who do not progress between levels (see, e.g., Jacobson and Truax, 1991). Our future analyses will include developing and testing such alternate metrics.

[4]  This document is undated and therefore it is possible that these score relationships rely upon outdated SAT, ACT, or other examinations.

[5]  These linkages make the assumption that a student who scores a grade equivalent of 10.8 on the CASAS exam would also score a 10.8 on the TABE. See Appendix B, especially Table B.1, for more information.

**Table 2.1**
**Categorization of TABE Scores**

| Category | TABE Grade Equivalent Score |
|---|---|
| Beginning adult basic education literacy (elementary school) | 0.0–1.9 |
| Beginning basic education (elementary school) | 2.0–3.9 |
| Low intermediate basic education (elementary school) | 4.0–5.9 |
| High intermediate basic education (middle school) | 6.0–8.9 |
| Low adult secondary education (early high school) | 9.0–10.9 |
| High adult secondary education (late high school) | 11.0–12.9 |

SOURCE: NRS (2015).

or higher grade equivalency score on the TABE have higher probabilities of passing the GED than individuals falling into lower Educational Functioning Levels.

Because of the detailed linkage between the NRS's Educational Functioning Levels and GED performance, we categorize grade equivalent TABE scores using the six Educational Functioning Levels identified by the NRS; these categories are listed in Table 2.1.

Beyond the world of adult education, there is a substantial amount of research documenting the use of standardized test scores to measure academic progress in the K–12 setting. The issue of standardized test scores has become considerably more prominent in K–12 public education since the passage of the No Child Left Behind (NCLB) law.[6] A key aspect of NCLB is the mandate of testing in reading and mathematics; students are tested yearly in grades 3–8 and once in high school and the focus of the testing is on determining whether students have achieved "grade-level proficiency" (or a *benchmark*) in each subject. Using a single benchmark metric of proficiency can be problematic; in particular, this metric could incentivize schools to focus only on students near the cut-off, which could result in resources being diverted away from students who are far below (or far above) the cut-off. In the case of the TABE, there is evidence that key benchmarks are meaningful; as discussed above, students who score at least 9.0 have a reasonably good probability of passing the GED, and those who score at least 11.0 have a very good possibility of passing the GED. This suggests that determining the number of cadets who achieve these grade levels could provide a meaningful metric of progress.

In short, combining benchmarks (indicators of the number who achieve key grade levels) with metrics of test score gains offers considerable advantage over other metrics. In the next section, we report both gain scores and the percentage of cadets who achieve key benchmarks (scoring at least the ninth-grade level and at least the eleventh-grade level on the final TABE). These results appear in Tables 2.4–2.9 (Figures 2.1–2.4 also provide relevant information). We also present the benchmarks for several subgroups. This provides a more complete picture of the progress that ChalleNGe cadets make in the classroom over the course of the program than what is provided by average gain scores. In future analysis, we will explore other options to avoid the documented problems with averaged gain scores as measured by GEs. We anticipate

---

[6]   NCLB, which was passed in 2001 and signed into law in early 2002, was a reauthorization of the Elementary and Secondary Education Act (ESEA) of 1965, and required that each state use standardized tests to measure students' progress against the state's curriculum standards. The 2016 reauthorization of ESEA, the Every Student Succeeds Act, includes the same stipulation. For more information, see "Every Student Succeeds Act (ESSA)" (n.d.).

eventually developing a metric for ChalleNGe that includes at least one benchmark TABE metric, as well as a metric of average improvement on the TABE.

## Cross-Site Metrics for the 2015 Classes

In this and the next section, we present information on the ChalleNGe classes that began in 2015 to document program progress and in support of the program's 2016 report to Congress. Table 2.2 provides a summary of ChalleNGe statistics across sites, while Tables 2.3–2.15 present information on the core components of ChalleNGe.

Table 2.2 provides a summary of the total numbers of applicants, enrollees, and graduates as well as indications of the number of academic credentials awarded and the hours (and dollar value) of community service provided by ChalleNGe cadets.

Tables 2.3–2.15 present data on the core components of ChalleNGe, focusing on metrics featured in previous annual reports. Information is provided on all sites and states (with some states containing multiple sites) to allow comparison of each component and metric. (See Table A.1 in Appendix A for the site abbreviations.) Appendix C presents detailed information individually for each ChalleNGe site. This information also serves to document site progress and support the annual report to Congress. The tables allow the reader to see all of a given site's information at once, thereby gaining a more detailed understanding of each site.

Individual data elements are occasionally left blank—this occurs for one of three reasons: the site only recently began operating and therefore lacks historical data; the site did not report the specific piece of data; or the site reported the data but the value appeared incorrect.[7] Finally, due to program timing, no site has yet collected 12-month postresidential placement data on cadets from Class 45. Therefore, we do not include this metric in any of the tables.

The tables are organized as follows:

- Credentials awarded (Table 2.3)
- TABE scores (Tables 2.4–2.9)
- Numbers of applicants and graduates (Table 2.1)
- Responsible citizenship (Tables 2.11 and 2.12)
- Community service (Table 2.13)
- Physical fitness (Tables 2.14 and 2.15).

Table 2.3 demonstrates a key aspect of site variation; the number and type of credentials awarded varies substantially across the sites. Following Table 2.15, we present basic analyses of the cadet-level data that sites reported. Each site provided information on initial and final TABE score for each cadet enrolled in the site, as well as cadet placement six months after completing the program. We focus on academic achievement and metrics based on the TABE categories presented above, as well as postresidential placements and cost per cadet. This analysis is presented at the aggregate level across all programs.

Some of the data in the tables above is quite similar to data in past annual reports. On many measures (such as test score gains, graduation rate, and placement rate), programs appear

---

[7]   In some cases, sites reported values that were out of the range of expected values. If these values could not be verified, they were considered missing. In a few cases, all values were identical; these too were treated as missing.

**Table 2.2**
**ChalleNGe Statistics, 1993–2015**

| ChalleNGe Statistics | 1993–2014[a] | 2015[b] | 1993–2015 |
|---|---|---|---|
| Applicants | 331,783 | 18,576 | 350,359 |
| Enrollees | 183,072 | 12,291 | 195,363 |
| Graduates | 136,693 | 9,230 | 145,923 |
| Academic credentials[c] | 88,614 | 4,104 | 92,718 |
| Hours of service to communities | 9,389,849 | 568,093 | 9,957,942 |
| Hours of service value | $183,963,578 | $13,035,555 | $196,999,134 |

[a] Historical information from ChalleNGe's 2015 annual report (National Guard Youth ChalleNGe, 2015); however, this report included Class 44 information from some sites. Therefore, 1993–2014 figures are adjusted to avoid double-counting.

[b] Information from the current data collection efforts (Classes 44 and 45), classes that began in 2015.

[c] Academic credentials reflect cadets who received either GED or HiSET or high school diploma (limited to one credential per cadet); therefore, the numbers reported in later tables do not sum to exactly the number reported here. Three sites (Florida, Maryland, and Puerto Rico) reported information only on graduates; therefore, the numbers of enrollees and applicants may represent a slight undercount. Additionally, programs may have reported total number of academic credentials for earlier classes; for classes beginning in 2015, we limited each cadet to a maximum of one academic credential.

**Table 2.3**
**Number of Credential(s) Awarded by Site (Classes 44 and 45)**

| Site | Residential Class 44 | | | | Residential Class 45 | | | |
|---|---|---|---|---|---|---|---|---|
| | Number of Graduates | GED or HiSET | High School (HS) Credits | HS Diploma | Number of Graduates | GED or HiSET | HS Credits | HS Diploma |
| All Sites | 4,500 | 1,477 | 2,014 | 676 | 4,730 | 1,797 | 1,926 | 880 |
| AK | 154 | 53 | 150 | 25 | 118 | 67 | 117 | 14 |
| AR | 96 | 24 | ~ | ~ | 97 | 15 | ~ | ~ |
| CA-LA | 193 | * | 193 | 19 | 184 | * | 184 | 18 |
| CA-SL | 191 | 5 | 189 | 54 | 208 | 4 | 208 | 60 |
| D.C. | 36 | * | ~ | 9 | 57 | * | ~ | 21 |
| FL | 165 | 65 | 165 | ~ | 167 | 73 | 34 | ~ |
| GA-FG | 192 | 79 | * | * | 182 | 113 | 7 | 5 |
| GA-FS | 172 | 112 | 18 | 18 | 214 | 148 | 33 | 32 |
| HI-BP | 81 | 81 | ~ | ~ | 129 | 129 | ~ | ~ |
| HI-HI | 60 | ~ | ~ | * | 70 | ~ | ~ | 70 |
| ID | 81 | 7 | 81 | 2 | 101 | 12 | 101 | 12 |
| IL | 195 | 113 | ~ | ~ | 174 | 85 | ~ | ~ |
| IN | 69 | 36 | ~ | ~ | 80 | 51 | ~ | ~ |
| KY-FK | 94 | 14 | 93 | ~ | 50 | 38 | 38 | ~ |
| KY-HN | 84 | 16 | 66 | ~ | 110 | 5 | 105 | ~ |
| LA-CB | 250 | 58 | ~ | ~ | 221 | 78 | ~ | ~ |
| LA-CM | 220 | 101 | ~ | ~ | 200 | 82 | ~ | ~ |
| LA-GL | 263 | 102 | ~ | ~ | 250 | 95 | ~ | ~ |
| MD | 84 | 34 | ~ | 27 | 107 | 62 | ~ | 62 |

Table 2.3—Continued

| Site | Residential Class 44 | | | | Residential Class 45 | | | |
|------|----------------------|--|--|--|----------------------|--|--|--|
| | Number of Graduates | GED or HiSET | High School (HS) Credits | HS Diploma | Number of Graduates | GED or HiSET | HS Credits | HS Diploma |
| MI | 106 | * | 106 | * | 107 | * | 107 | * |
| MS | 168 | 94 | ~ | 94 | 206 | 129 | ~ | 129 |
| MT | 84 | 42 | 84 | ~ | 74 | 34 | 74 | ~ |
| NC-NL | ^ | ^ | ^ | ^ | 50 | 27 | ~ | ~ |
| NC-S | 106 | 48 | ~ | ~ | 140 | 47 | ~ | ~ |
| NJ | 100 | 44 | ~ | 44 | 99 | 44 | ~ | 44 |
| NM | 80 | 31 | ~ | ~ | 94 | 77 | ~ | ~ |
| OK | 128 | 23 | 128 | ~ | 101 | 12 | 101 | 3 |
| OR | 125 | 6 | 125 | 28 | 134 | 4 | 134 | 22 |
| PR | 225 | ~ | 225 | 225 | 225 | ~ | 225 | 225 |
| SC | 94 | 21 | 2 | 1 | 103 | 47 | 1 | * |
| TX-E | ^ | ^ | ^ | ^ | 51 | 24 | 50 | 12 |
| TX-W | 93 | 43 | 93 | 9 | 74 | 42 | 74 | 4 |
| VA | 66 | 10 | 11 | ~ | 91 | 28 | 25 | ~ |
| WA | 140 | ~ | 140 | ~ | 152 | ~ | 152 | ~ |
| WI | 103 | 53 | * | * | 100 | 48 | * | * |
| WV | 138 | 116 | 137 | 116 | 152 | 145 | 151 | 145 |
| WY | 64 | 46 | 8 | 4 | 58 | 32 | 5 | 2 |

NOTES: Information in the table includes Classes 44 and 45 (which generally began and ended in 2015). In this table only, blanks occur either because the site is newly operational, did not report the data, or does not award the specific credential. Credentials awarded include those awarded during the course of the ChalleNGe Residential Phase; some programs may also have included additional credentials awarded soon after the end of the Residential Phase. We counted only a single credential per cadet.

* Did not report, or inconsistent with other reported information.

^ Newly operational.

~ Does not award.

to be performing (on average) very much like past programs. However, the tables above also demonstrate the substantial variation that exists between programs. For example, programs vary dramatically in terms of applicants, ratio of entrants to applicants, meeting program target in terms of graduates, ratio of graduates to entrants, and average initial test scores. Analyses to determine the extent to which ChalleNGe policies versus, for example, state-level differences drive program-level variation will be a topic of future reports.

## Cross-Program Analyses

We now present basic analyses of the cadet-level data collected from each site and summarized in the previous section, focusing on academic achievement and metrics based on the TABE categories presented in Table 2.1. When reporting TABE scores, we report TABE (Total) Battery grade equivalent scores. This metric is formed by combining performance in both math

**Table 2.4**
**Average TABE Math Score and Gain by Site (Classes 44 and 45)**

| Site | Residential Class 44 | | | Residential Class 45 | | |
|---|---|---|---|---|---|---|
| | Pre-TABE | Post-TABE | Gain (+/−) | Pre-TABE | Post-TABE | Gain (+/−) |
| All Sites | 6.4 | 8.3 | 1.9 | 6.3 | 8.4 | 2.1 |
| AK | 8.2 | 9.7 | 1.5 | 7.8 | 9.8 | 2.0 |
| AR | 7.4 | 8.8 | 1.4 | 7.0 | 8.5 | 1.5 |
| CA-LA | 5.5 | 4.8 | −0.7 | 5.8 | 7.8 | 2.0 |
| CA-SL | 6.8 | 8.2 | 1.4 | 7.3 | 8.7 | 1.4 |
| D.C. | 5.6 | 6.7 | 1.1 | 5.7 | 7.8 | 2.1 |
| FL | 6.4 | 7.9 | 1.5 | 6.5 | 9.4 | 2.9 |
| GA-FG | * | * | * | 4.2 | 7.2 | 3.0 |
| GA-FS | 6.9 | 9.9 | 3.0 | 7.2 | 10.3 | 3.1 |
| HI-BP | 5.5 | 7.9 | 2.4 | 5.0 | 5.3 | 0.3 |
| HI-HI | 5.4 | 6.3 | 0.9 | 5.2 | 6.3 | 1.1 |
| ID | 8.1 | 9.9 | 1.8 | 7.3 | 9.6 | 2.3 |
| IL | 6.4 | 9.0 | 2.6 | 6.0 | 9.8 | 3.8 |
| IN | 6.9 | 8.5 | 1.6 | 6.8 | 8.1 | 1.3 |
| KY-FK | 6.5 | 9.3 | 2.8 | 5.7 | 9.0 | 3.3 |
| KY-HN | 5.3 | 4.9 | −0.4 | 5.6 | 4.4 | −1.2 |
| LA-CB | 6.3 | 9.9 | 3.6 | 6.6 | 10.0 | 3.4 |
| LA-CM | 6.5 | 9.6 | 3.1 | 6.5 | 9.3 | 2.8 |
| LA-GL | 6.7 | 7.3 | 0.6 | 6.1 | 6.8 | 0.7 |
| MD | 6.0 | 9.1 | 3.1 | 5.6 | 8.6 | 3.0 |
| MI | 6.9 | 8.2 | 1.3 | 7.2 | 7.8 | 0.6 |
| MS | 5.3 | 9.9 | 4.6 | 5.5 | 10.3 | 4.8 |
| MT | 8.2 | 9.9 | 1.7 | 8.5 | 9.7 | 1.2 |
| NC-NL | ^ | ^ | ^ | 6.4 | 8.5 | 2.1 |
| NC-S | 6.3 | 8.7 | 2.4 | 6.3 | 8.3 | 2.0 |
| NJ | 5.9 | 9.2 | 3.3 | 6.7 | 9.2 | 2.5 |
| NM | 5.9 | 8.4 | 2.5 | 6.1 | 8.1 | 2.0 |
| OK | 7.3 | 8.5 | 1.2 | 7.4 | 7.8 | 0.4 |
| OR | 6.7 | 7.4 | 0.7 | 6.9 | 7.9 | 1.0 |
| PR | 3.9 | 6.3 | 2.4 | 3.9 | 6.1 | 2.2 |
| SC | 6.2 | 6.8 | 0.6 | 7.0 | 7.6 | 0.6 |
| TX-E | ^ | ^ | ^ | 7.7 | 8.4 | 0.7 |
| TX-W | 7.3 | 5.9 | −1.4 | 7.3 | 9.0 | 1.7 |
| VA | 6.4 | 7.6 | 1.2 | 6.0 | 7.4 | 1.4 |

**Table 2.4—Continued**

| Site | Residential Class 44 | | | Residential Class 45 | | |
|------|----------|-----------|-----------|----------|-----------|-----------|
| | Pre-TABE | Post-TABE | Gain (+/−) | Pre-TABE | Post-TABE | Gain (+/−) |
| WA | 6.4 | 8.7 | 2.3 | 6.1 | 9.1 | 3.0 |
| WI | 7.9 | 8.3 | 0.4 | 8.1 | 8.8 | 0.7 |
| WV | 6.0 | 8.2 | 2.2 | 6.7 | 8.4 | 1.7 |
| WY | 8.1 | 9.0 | 0.9 | 7.2 | 7.5 | 0.3 |

NOTE: Information in the table includes Classes 44 and 45 (which generally began and ended in 2015). Blanks in the table occur due to new sites (which have little historical data), or in cases where data were not reported or did not appear correct.

* Did not report.

^ Newly operational.

**Table 2.5**
**Average TABE Battery Score and Gain by Site (Classes 44 and 45)**

| Site | Residential Class 44 | | | Residential Class 45 | | |
|------|----------|-----------|-----------|----------|-----------|-----------|
| | Pre-TABE | Post-TABE | Gain (+/−) | Pre-TABE | Post-TABE | Gain (+/−) |
| All Sites | 7.0 | 8.8 | 1.8 | 6.8 | 8.7 | 1.9 |
| AK | 8.3 | 9.6 | 1.3 | 8.1 | 9.7 | 1.6 |
| AR | 8.2 | 9.3 | 1.1 | 7.9 | 9.3 | 1.4 |
| CA-LA | 8.1 | 8.7 | 0.6 | 5.7 | 8.6 | 2.9 |
| CA-SL | 7.1 | 8.6 | 1.5 | 7.5 | 8.9 | 1.4 |
| D.C. | 5.2 | 6.9 | 1.7 | 5.5 | 7.1 | 1.6 |
| FL | 6.7 | 8.4 | 1.7 | 6.9 | 10.1 | 3.2 |
| GA-FG | * | * | * | 5.1 | 7.8 | 2.7 |
| GA-FS | 7.3 | 10.0 | 2.7 | 7.1 | 8.4 | 1.3 |
| HI-BP | 5.3 | 8.0 | 2.7 | 7.2 | 7.9 | 0.7 |
| HI-HI | 5.5 | 6.0 | 0.5 | 4.6 | 5.6 | 1.0 |
| ID | 8.3 | 10.3 | 2.0 | 7.4 | 9.9 | 2.5 |
| IL | 8.1 | 9.2 | 1.1 | 6.4 | 8.8 | 2.4 |
| IN | 6.9 | 8.5 | 1.6 | 6.9 | 8.6 | 1.7 |
| KY-FK | 6.6 | 7.6 | 1.0 | 4.9 | 7.4 | 2.5 |
| KY-HN | 4.9 | 5.0 | 0.1 | 4.7 | 3.9 | −0.8 |
| LA-CB | 6.9 | 9.9 | 3.0 | 7.1 | 9.9 | 2.8 |
| LA-CM | 7.1 | 9.5 | 2.4 | 6.9 | 9.1 | 2.2 |
| LA-GL | 9.4 | 9.6 | 0.2 | 8.6 | 8.9 | 0.3 |
| MD | 6.0 | 9.9 | 3.9 | 5.9 | 9.1 | 3.2 |
| MI | 7.2 | 8.0 | 0.8 | * | * | * |
| MS | 6.0 | 10.2 | 4.2 | 6.2 | 10.2 | 4.0 |
| MT | 7.6 | 8.9 | 1.3 | 8.1 | 8.8 | 0.7 |
| NC-NL | ^ | ^ | ^ | 6.3 | 8.7 | 2.4 |
| NC-S | 6.2 | 9.2 | 3.0 | 6.1 | 8.5 | 2.4 |

**Table 2.5—Continued**

| Site | Residential Class 44 | | | Residential Class 45 | | |
|------|------------|-------------|-----------|------------|-------------|-----------|
| | Pre-TABE | Post-TABE | Gain (+/−) | Pre-TABE | Post-TABE | Gain (+/−) |
| NJ | 6.4 | 9.2 | 2.8 | 6.9 | 9.2 | 2.3 |
| NM | 6.2 | 8.5 | 2.3 | 6.0 | 7.9 | 1.9 |
| OK | 7.0 | 8.4 | 1.4 | 7.5 | 7.5 | 0.0 |
| OR | 8.5 | 9.1 | 0.6 | 8.7 | 9.3 | 0.6 |
| PR | 4.3 | 7.7 | 3.4 | 4.3 | 7.6 | 3.3 |
| SC | 6.1 | 6.4 | 0.3 | 7.2 | 7.6 | 0.4 |
| TX-E | ^ | ^ | ^ | 7.7 | 8.3 | 0.6 |
| TX-W | 7.4 | 7.3 | −0.1 | 7.1 | 8.4 | 1.3 |
| VA | 6.4 | 8.0 | 1.6 | 6.3 | 7.8 | 1.5 |
| WA | 6.8 | 8.9 | 2.1 | 6.7 | 9.2 | 2.5 |
| WI | 8.3 | 8.3 | 0.0 | 7.9 | 9.0 | 1.1 |
| WV | 6.7 | 9.8 | 3.1 | 6.9 | 9.3 | 2.4 |
| WY | 8.8 | 9.6 | 0.8 | 8.5 | 8.8 | 0.3 |

NOTE: Information in the table includes Classes 44 and 45 (which generally began and ended in 2015). Blanks in the table occur due to new sites (which have little historical data), or in cases where data were not reported or did not appear correct.

* Did not report, or inconsistent with other reported information.

^ Newly operational.

**Table 2.6**
**Distribution of Pre- and Post-TABE Math Scores by Site (Class 44)**

| Site | Pre-TABE | | | Post-TABE | | |
|------|------------------------------|------------------------------|----------------------------|------------------------------|------------------------------|----------------------------|
| | Elementary (Grades 1–6) | Middle School (Grades 7–8) | High School (Grades 9–12) | Elementary (Grades 1–6) | Middle School (Grades 7–8) | High School (Grades 9–12) |
| All Sites | 2,270 | 1,144 | 851 | 1,143 | 1,358 | 1,740 |
| AK | 53 | 41 | 60 | 17 | 44 | 93 |
| AR | 40 | 29 | 26 | 17 | 37 | 41 |
| CA-LA | 123 | 49 | 19 | 138 | 40 | 12 |
| CA-SL | 97 | 45 | 48 | 48 | 69 | 72 |
| D.C. | 20 | 13 | 1 | 11 | 20 | 3 |
| FL | 84 | 54 | 27 | 40 | 72 | 53 |
| GA-FG | * | * | * | * | * | * |
| GA-FS | 70 | 72 | 30 | 9 | 51 | 110 |
| HI-BP | 53 | 15 | 12 | 27 | 24 | 29 |
| HI-HI | 39 | 11 | 9 | 32 | 13 | 14 |
| ID | 24 | 25 | 32 | 4 | 29 | 48 |
| IL | 103 | 56 | 30 | 39 | 52 | 94 |
| IN | 35 | 14 | 20 | 17 | 21 | 31 |
| KY-FK | 46 | 33 | 15 | 17 | 28 | 49 |

**Table 2.6—Continued**

| Site | Pre-TABE | | | Post-TABE | | |
|------|----------|--|--|-----------|--|--|
| | Elementary (Grades 1–6) | Middle School (Grades 7–8) | High School (Grades 9–12) | Elementary (Grades 1–6) | Middle School (Grades 7–8) | High School (Grades 9–12) |
| KY-HN | 50 | 19 | 7 | 53 | 15 | 7 |
| LA-CB | 140 | 66 | 43 | 23 | 69 | 156 |
| LA-CM | 120 | 44 | 54 | 31 | 61 | 128 |
| LA-GL | 123 | 74 | 56 | 94 | 85 | 75 |
| MD | 55 | 15 | 14 | 8 | 34 | 41 |
| MI | 51 | 28 | 27 | 26 | 37 | 43 |
| MS | 118 | 37 | 13 | 9 | 56 | 103 |
| MT | 24 | 22 | 38 | 6 | 17 | 51 |
| NC-NL | ^ | ^ | ^ | ^ | ^ | ^ |
| NC-S | 56 | 31 | 18 | 32 | 28 | 46 |
| NJ | 51 | 31 | 18 | 19 | 32 | 49 |
| NM | 46 | 25 | 9 | 16 | 29 | 35 |
| OK | 52 | 42 | 34 | 30 | 38 | 60 |
| OR | 59 | 39 | 27 | 51 | 32 | 42 |
| PR | 210 | 14 | 1 | 121 | 75 | 29 |
| SC | 57 | 20 | 17 | 41 | 25 | 21 |
| TX-E | ^ | ^ | ^ | ^ | ^ | ^ |
| TX-W | 40 | 21 | 29 | 39 | 50 | 0 |
| VA | 33 | 19 | 12 | 24 | 18 | 23 |
| WA | 71 | 44 | 25 | 26 | 56 | 58 |
| WI | 33 | 35 | 35 | 26 | 36 | 41 |
| WV | 78 | 38 | 22 | 40 | 46 | 50 |
| WY | 16 | 23 | 23 | 12 | 19 | 33 |

NOTE: Information in the table includes Classes 44 and 45 (which generally began and ended in 2015). Blanks in the table occur due to new sites (which have little historical data), or in cases when data were not reported or did not appear correct.

* Did not report.

^ Newly operational.

**Table 2.7**
**Distribution of Pre- and Post-TABE Math Scores by Site (Class 45)**

| Site | Pre-TABE | | | Post-TABE | | |
|------|----------|--|--|-----------|--|--|
| | Elementary (Grades 1–6) | Middle School (Grades 7–8) | High School (Grades 9–12) | Elementary (Grades 1–6) | Middle School (Grades 7–8) | High School (Grades 9–12) |
| All Sites | 2,461 | 1,192 | 912 | 1,232 | 1,355 | 1,930 |
| AK | 41 | 32 | 45 | 15 | 32 | 71 |
| AR | 47 | 23 | 26 | 27 | 24 | 46 |
| CA-LA | 106 | 48 | 26 | 56 | 65 | 60 |
| CA-SL | 95 | 46 | 67 | 37 | 77 | 94 |

**Table 2.7—Continued**

| Site | Pre-TABE | | | Post-TABE | | |
|------|----------|--|--|-----------|--|--|
| | Elementary (Grades 1–6) | Middle School (Grades 7–8) | High School (Grades 9–12) | Elementary (Grades 1–6) | Middle School (Grades 7–8) | High School (Grades 9–12) |
| D.C. | 22 | 10 | 2 | 12 | 11 | 12 |
| FL | 79 | 56 | 32 | 21 | 41 | 105 |
| GA-FG | 115 | 15 | 8 | 45 | 48 | 27 |
| GA-FS | 80 | 81 | 52 | 11 | 56 | 146 |
| HI-BP | 90 | 30 | 9 | 89 | 26 | 13 |
| HI-HI | 49 | 19 | 2 | 33 | 33 | 4 |
| ID | 42 | 28 | 31 | 14 | 29 | 58 |
| IL | 62 | 23 | 16 | 8 | 30 | 63 |
| IN | 42 | 22 | 16 | 24 | 28 | 28 |
| KY-FK | 32 | 11 | 5 | 7 | 14 | 28 |
| KY-HN | 60 | 28 | 15 | 68 | 12 | 8 |
| LA-CB | 114 | 54 | 53 | 29 | 43 | 149 |
| LA-CM | 101 | 56 | 42 | 26 | 75 | 99 |
| LA-GL | 136 | 81 | 31 | 111 | 76 | 61 |
| MD | 71 | 24 | 12 | 23 | 36 | 47 |
| MI | 45 | 27 | 35 | 37 | 31 | 39 |
| MS | 133 | 52 | 21 | 10 | 57 | 139 |
| MT | 15 | 21 | 37 | 7 | 14 | 44 |
| NC-NL | 27 | 13 | 7 | 15 | 11 | 21 |
| NC-S | 77 | 39 | 24 | 40 | 44 | 56 |
| NJ | 44 | 30 | 24 | 14 | 36 | 49 |
| NM | 52 | 27 | 15 | 26 | 32 | 36 |
| OK | 43 | 26 | 32 | 39 | 25 | 37 |
| OR | 60 | 37 | 36 | 48 | 39 | 46 |
| PR | 206 | 16 | 2 | 130 | 72 | 23 |
| SC | 46 | 31 | 26 | 41 | 23 | 33 |
| TX-E | 20 | 14 | 17 | 12 | 19 | 20 |
| TX-W | 28 | 23 | 23 | 18 | 15 | 40 |
| VA | 55 | 20 | 16 | 33 | 26 | 27 |
| WA | 91 | 36 | 25 | 22 | 56 | 74 |
| WI | 39 | 21 | 40 | 24 | 27 | 49 |
| WV | 71 | 53 | 28 | 39 | 52 | 61 |
| WY | 25 | 19 | 14 | 21 | 20 | 17 |

NOTE: Information in the table includes Class 45.

**Table 2.8**
**Distribution of Pre- and Post-TABE Battery Scores by Site (Class 44)**

| Site | Pre-TABE | | | Post-TABE | | |
|---|---|---|---|---|---|---|
| | Elementary (Grades 1–6) | Middle School (Grades 7–8) | High School (Grades 9–12) | Elementary (Grades 1–6) | Middle School (Grades 7–8) | High School (Grades 9–12) |
| All Sites | 1,854 | 1,233 | 1,181 | 858 | 1,188 | 2,196 |
| AK | 45 | 45 | 64 | 24 | 33 | 97 |
| AR | 22 | 35 | 38 | 10 | 30 | 55 |
| CA-LA | 55 | 68 | 67 | 42 | 56 | 92 |
| CA-SL | 70 | 75 | 46 | 28 | 78 | 83 |
| D.C. | 23 | 10 | 1 | 12 | 17 | 5 |
| FL | 71 | 62 | 32 | 27 | 70 | 68 |
| GA-FG | * | * | * | * | * | * |
| GA-FS | 65 | 58 | 49 | 16 | 41 | 114 |
| HI-BP | 57 | 13 | 11 | 20 | 34 | 26 |
| HI-HI | 38 | 12 | 9 | 33 | 9 | 17 |
| ID | 25 | 24 | 32 | 7 | 16 | 58 |
| IL | 57 | 63 | 69 | 29 | 45 | 112 |
| IN | 32 | 16 | 21 | 21 | 15 | 33 |
| KY-FK | 48 | 25 | 21 | 37 | 21 | 35 |
| KY-HN | 57 | 17 | 8 | 59 | 10 | 12 |
| LA-CB | 114 | 76 | 60 | 32 | 51 | 167 |
| LA-CM | 101 | 51 | 66 | 37 | 52 | 131 |
| LA-GL | 27 | 76 | 145 | 34 | 56 | 158 |
| MD | 49 | 25 | 10 | 4 | 22 | 57 |
| MI | 42 | 36 | 28 | 29 | 40 | 37 |
| MS | 97 | 49 | 22 | 9 | 41 | 118 |
| MT | 31 | 24 | 29 | 15 | 17 | 42 |
| NC-NL | ^ | ^ | ^ | ^ | ^ | ^ |
| NC-S | 55 | 31 | 20 | 24 | 25 | 57 |
| NJ | 52 | 27 | 21 | 17 | 27 | 56 |
| NM | 44 | 22 | 13 | 14 | 29 | 36 |
| OK | 59 | 32 | 37 | 28 | 42 | 58 |
| OR | 26 | 44 | 55 | 24 | 33 | 68 |
| PR | 199 | 20 | 6 | 63 | 78 | 84 |
| SC | 57 | 22 | 15 | 47 | 18 | 22 |
| TX-E | ^ | ^ | ^ | ^ | ^ | ^ |
| TX-W | 35 | 25 | 29 | 25 | 38 | 26 |
| VA | 33 | 16 | 15 | 17 | 21 | 26 |

**Table 2.8—Continued**

|  | Pre-TABE | | | Post-TABE | | |
|---|---|---|---|---|---|---|
| Site | Elementary (Grades 1–6) | Middle School (Grades 7–8) | High School (Grades 9–12) | Elementary (Grades 1–6) | Middle School (Grades 7–8) | High School (Grades 9–12) |
| WA | 65 | 39 | 36 | 20 | 46 | 74 |
| WI | 25 | 33 | 45 | 34 | 20 | 49 |
| WV | 67 | 41 | 30 | 12 | 41 | 83 |
| WY | 11 | 21 | 31 | 8 | 16 | 40 |

NOTE: Information in the table includes Class 44. Blanks in the table occur due to new sites (which have little historical data), or in cases where data were not reported or did not appear correct.

* Did not report.

^ Newly operational.

**Table 2.9**
**Distribution of Pre- and Post-TABE Battery Scores by Site (Class 45)**

|  | Pre-TABE | | | Post-TABE | | |
|---|---|---|---|---|---|---|
| Site | Elementary (Grades 1–6) | Middle School (Grades 7–8) | High School (Grades 9–12) | Elementary (Grades 1–6) | Middle School (Grades 7–8) | High School (Grades 9–12) |
| All Sites | 2,127 | 1,194 | 1,145 | 968 | 1,287 | 2,171 |
| AK | 36 | 32 | 49 | 14 | 28 | 76 |
| AR | 30 | 27 | 40 | 14 | 27 | 56 |
| CA-LA | 112 | 42 | 27 | 36 | 55 | 90 |
| CA-SL | 72 | 72 | 64 | 24 | 83 | 101 |
| D.C. | 19 | 12 | 4 | 14 | 12 | 9 |
| FL | 75 | 46 | 46 | 11 | 34 | 122 |
| GA-FG | 99 | 25 | 14 | 32 | 46 | 42 |
| GA-FS | 92 | 61 | 60 | 59 | 51 | 103 |
| HI-BP | 49 | 50 | 29 | 36 | 45 | 47 |
| HI-HI | 56 | 10 | 4 | 39 | 27 | 3 |
| ID | 44 | 22 | 35 | 10 | 28 | 63 |
| IL | 51 | 33 | 18 | 11 | 40 | 51 |
| IN | 38 | 20 | 22 | 19 | 23 | 38 |
| KY-FK | 37 | 5 | 5 | 15 | 19 | 15 |
| KY-HN | 82 | 15 | 10 | 82 | 16 | 5 |
| LA-CB | 91 | 67 | 63 | 33 | 47 | 140 |
| LA-CM | 94 | 55 | 51 | 40 | 56 | 104 |
| LA-GL | 61 | 68 | 119 | 53 | 63 | 132 |
| MD | 64 | 30 | 13 | 14 | 36 | 55 |
| MI | * | * | * | * | * | * |
| MS | 118 | 54 | 34 | 10 | 53 | 143 |
| MT | 22 | 21 | 30 | 12 | 22 | 31 |
| NC-NL | 25 | 15 | 8 | 16 | 10 | 23 |

Table 2.9—Continued

| | Pre-TABE | | | Post-TABE | | |
|---|---|---|---|---|---|---|
| Site | Elementary (Grades 1–6) | Middle School (Grades 7–8) | High School (Grades 9–12) | Elementary (Grades 1–6) | Middle School (Grades 7–8) | High School (Grades 9–12) |
| NC-S | 81 | 36 | 23 | 34 | 46 | 60 |
| NJ | 42 | 31 | 25 | 17 | 25 | 57 |
| NM | 57 | 21 | 16 | 23 | 39 | 31 |
| OK | 41 | 22 | 38 | 42 | 22 | 37 |
| OR | 25 | 46 | 63 | 15 | 45 | 74 |
| PR | 182 | 32 | 11 | 78 | 67 | 80 |
| SC | 46 | 26 | 31 | 40 | 20 | 38 |
| TX-E | 19 | 14 | 18 | 12 | 15 | 23 |
| TX-W | 31 | 20 | 23 | 17 | 20 | 37 |
| VA | 49 | 26 | 16 | 21 | 36 | 29 |
| WA | 68 | 46 | 38 | 26 | 38 | 88 |
| WI | 36 | 28 | 36 | 19 | 34 | 47 |
| WV | 66 | 47 | 39 | 18 | 40 | 94 |
| WY | 17 | 17 | 23 | 12 | 19 | 27 |

NOTE: Information in the table includes Class 45. Blanks in the table occur due to new sites (which have little historical data), or in cases where data were not reported or did not appear correct.

* Did not report.

Table 2.10
Applicants and Graduates (Classes 44 and 45)

| | Residential Class 44 | | | Residential Class 45 | | |
|---|---|---|---|---|---|---|
| Site | Target | Applied | Graduates | Target | Applied | Graduates |
| All Sites | 4,872 | 8,908 | 4,500 | 5,125 | 9,668 | 4,730 |
| AK | 144 | 232 | 154 | 144 | 184 | 118 |
| AR | 100 | 195 | 96 | 100 | 233 | 97 |
| CA-LA | 180 | 255 | 193 | 180 | 208 | 184 |
| CA-SL | 180 | 247 | 191 | 185 | 304 | 208 |
| D.C. | 100 | 65 | 36 | 100 | 81 | 57 |
| FL | 150 | 245 | 165 | 150 | 265 | 167 |
| GA-FG | 212 | 533 | 192 | 213 | 389 | 182 |
| GA-FS | 213 | 374 | 172 | 212 | 395 | 214 |
| HI-BP | 100 | 195 | 81 | 100 | 224 | 129 |
| HI-HI | 100 | 108 | 60 | 100 | 112 | 70 |
| ID | 100 | 108 | 81 | 100 | 127 | 101 |
| IL | 300 | 476 | 195 | 300 | 420 | 174 |
| IN | 100 | 130 | 69 | 100 | 172 | 80 |
| KY-FK | 100 | 136 | 94 | 100 | 118 | 50 |
| KY-HN | 100 | 118 | 84 | 100 | 187 | 110 |

**Table 2.12**
**Core Component Completion—Responsible Citizenship (Class 45)**

| Site | Eligible to Vote | Registered to Vote | % Eligible Who Registered | Eligible for Selective Service | Registered for Selective Service | % Eligible Who Registered |
|---|---|---|---|---|---|---|
| All Sites | 1,205 | 1,100 | 91% | 1,454 | 1,402 | 96% |
| AK | 32 | 32 | 100% | 26 | 26 | 100% |
| AR | 28 | 28 | 100% | 55 | 55 | 100% |
| CA-LA | 34 | 34 | 100% | 34 | 34 | 100% |
| CA-SL | 43 | 43 | 100% | 43 | 43 | 100% |
| D.C. | 8 | 8 | 100% | 7 | 7 | 100% |
| FL | 54 | 54 | 100% | 43 | 43 | 100% |
| GA-FG | 62 | 62 | 100% | 83 | 83 | 100% |
| GA-FS | 67 | 67 | 100% | 57 | 57 | 100% |
| HI-BP | 32 | 32 | 100% | 20 | 20 | 100% |
| HI-HI | 11 | 11 | 100% | 38 | 38 | 100% |
| ID | 19 | 19 | 100% | 31 | 31 | 100% |
| IL | 38 | 38 | 100% | 27 | 27 | 100% |
| IN | 15 | 15 | 100% | 32 | 32 | 100% |
| KY-FK | 7 | 7 | 100% | 13 | 13 | 100% |
| KY-HN | 20 | 20 | 100% | 20 | 20 | 100% |
| LA-CB | 35 | 32 | 91% | 108 | 108 | 100% |
| LA-CM | 30 | 30 | 100% | 54 | 44 | 81% |
| LA-GL | 49 | 49 | 100% | 34 | 34 | 100% |
| MD | 42 | 42 | 100% | 46 | 46 | 100% |
| MI | 33 | 0 | 0% | 33 | 9 | 27% |
| MS | 51 | 51 | 100% | 81 | 81 | 100% |
| MT | 15 | 0 | 0% | 25 | 25 | 100% |
| NC-NL | 15 | 15 | 100% | 12 | 12 | 100% |
| NC-S | 33 | 33 | 100% | 30 | 30 | 100% |
| NJ | 29 | 29 | 100% | 21 | 21 | 100% |
| NM | 73 | 19 | 26% | 42 | 34 | 81% |
| OK | 9 | 9 | 100% | 25 | 25 | 100% |
| OR | 42 | 42 | 100% | 83 | 83 | 100% |
| PR | 67 | 67 | 100% | 53 | 53 | 100% |
| SC | 24 | 24 | 100% | 23 | 23 | 100% |
| TX-E | 15 | 15 | 100% | 15 | 15 | 100% |
| TX-W | 29 | 29 | 100% | 41 | 41 | 100% |
| VA | 22 | 22 | 100% | 47 | 46 | 98% |
| WA | 48 | 48 | 100% | 51 | 51 | 100% |
| WI | 31 | 31 | 100% | 55 | 55 | 100% |
| WV | 34 | 34 | 100% | 28 | 28 | 100% |
| WY | 9 | 9 | 100% | 18 | 9 | 50% |

NOTE: Information in the table includes Class 45.

**Table 2.13**
**Core Component Completion—Community Service (Classes 44 and 45)**

| Site | Residential Class 44 | | | Residential Class 45 | | |
|---|---|---|---|---|---|---|
| | Total Hours, Comm. Svc. | Dollar Value/Hr | Total Community Service Contribution | Total Hours Comm. Svc. | Dollar Value/Hr | Total Community Service Contribution |
| All Sites | 278,940 | | $6,411,472 | 289,153 | | $6,624,083 |
| AK | 10,984 | $27.51 | $302,170 | 6,809 | $27.51 | $187,316 |
| AR | 7,213 | $19.14 | $138,057 | 7,071 | $19.14 | $135,339 |
| CA-LA | 8,695 | $27.59 | $239,895 | 8,423 | $27.59 | $232,377 |
| CA-SL | 20,024 | $27.59 | $552,462 | 13,056 | $27.59 | $360,215 |
| D.C. | 2,150 | $38.77 | $83,356 | 1,505 | $38.77 | $58,349 |
| FL | 8,248 | $22.08 | $182,116 | 8,038 | $22.08 | $177,479 |
| GA-FG | 11,342 | $23.80 | $269,940 | 10,159 | $23.80 | $241,784 |
| GA-FS | 10,258 | $23.80 | $244,140 | 12,130 | $23.80 | $288,682 |
| HI-BP | 9,212 | $23.33 | $214,904 | 15,401 | $23.33 | $359,294 |
| HI-HI | 6,708 | $23.33 | $156,498 | 9,402 | $23.33 | $219,349 |
| ID | 4,904 | $20.97 | $102,837 | 4,474 | $20.97 | $93,809 |
| IL | 11,896 | $25.34 | $301,445 | 11,854 | $25.34 | $300,380 |
| IN | 4,059 | $22.69 | $92,087 | 4,848 | $22.69 | $110,001 |
| KY-FK | 3,140 | $21.16 | $66,432 | 3,940 | $21.16 | $83,370 |
| KY-HN | 5,761 | $21.16 | $121,903 | 7,270 | $21.16 | $153,833 |
| LA-CB | 11,888 | $22.67 | $269,501 | 10,145 | $22.67 | $229,987 |
| LA-CM | 9,352 | $22.67 | $212,010 | 10,187 | $22.67 | $230,939 |
| LA-GL | 16,411 | $22.67 | $372,037 | 19,807 | $22.67 | $449,025 |
| MD | 4,049 | $26.64 | $107,865 | 4,855 | $26.64 | $129,326 |
| MI | 4,869 | $23.54 | $114,616 | 3,992 | $23.54 | $93,960 |
| MS | 14,490 | $19.51 | $282,690 | 13,247 | $19.51 | $258,439 |
| MT | 4,270 | $20.44 | $87,284 | 4,338 | $20.44 | $88,677 |
| NC-NL | ^ | ^ | ^ | 2,000 | $21.88 | $43,760 |
| NC-S | 8,303 | $21.88 | $181,677 | 9,765 | $21.88 | $213,658 |
| NJ | 4,350 | $26.70 | $116,145 | 5,473 | $26.70 | $146,116 |
| NM | 6,491 | $19.91 | $129,226 | 4,322 | $19.91 | $86,051 |
| OK | 8,911 | $21.50 | $191,581 | 6,824 | $21.50 | $146,716 |
| OR | 11,043 | $22.75 | $251,223 | 12,773 | $22.75 | $290,574 |
| PR | 13,360 | $11.39 | $152,170 | 11,504 | $11.39 | $131,031 |
| SC | 5,246 | $21.14 | $110,900 | 5,586 | $21.14 | $118,088 |
| TX-E | ^ | ^ | ^ | 1,811 | $25.11 | $45,474 |
| TX-W | 4,093 | $25.11 | $102,763 | 3,150 | $25.11 | $79,084 |
| VA | 2,716 | $26.09 | $70,860 | 7,162 | $26.09 | $186,857 |

**Table 2.13—Continued**

| Site | Residential Class 44 | | | Residential Class 45 | | |
|---|---|---|---|---|---|---|
| | Total Hours, Comm. Svc. | Dollar Value/Hr | Total Community Service Contribution | Total Hours Comm. Svc. | Dollar Value/Hr | Total Community Service Contribution |
| WA | 7,810 | $28.99 | $226,397 | 7,292 | $28.99 | $211,389 |
| WI | 7,241 | $22.48 | $162,783 | 7,539 | $22.48 | $169,471 |
| WV | 6,477 | $20.47 | $132,574 | 10,126 | $20.47 | $207,269 |
| WY | 2,980 | $23.13 | $68,927 | 2,880 | $23.13 | $66,614 |

NOTE: Information in the table includes Classes 44 and 45. Blanks in the table occur due to new sites. The total hours of community service (Comm. Svc.) was reported by each site in the program survey. The figures for dollar value per hour were obtained from published figures at the state level for 2015 and are available online at the Independent Sector (http://www.independentsector.org/resource/the-value-of-volunteer-time/). The ChalleNGe program utilized the same source of information for the 2015 Performance and Accountability Highlights report.

^Newly operational.

**Table 2.14**
**Residential Performance—Physical Fitness as Measured by the Average Number Completed and Time for Cadets per Site (Class 44)**

| Site | Curl-Ups | | Push-Ups | | 1-Mile Run | |
|---|---|---|---|---|---|---|
| | Initial | Final | Initial | Final | Initial | Final |
| All Sites | 33.1 | 48.9 | 24.3 | 41.0 | 10:09 | 08:29 |
| AK | 37.9 | 46.6 | * | * | 10:27 | 07:47 |
| AR | 29.3 | 29.6 | 22.8 | 44.3 | 11:01 | 11:20 |
| CA-LA | 30.4 | 44.8 | 25.1 | 56.8 | 08:19 | 07:02 |
| CA-SL | 30.6 | 43.7 | 20.3 | 36.2 | 09:21 | 07:33 |
| D.C. | 26.5 | 34.2 | 16.1 | 28.2 | 10:50 | 09:15 |
| FL | 35.9 | 71.8 | 19.9 | 36.0 | 10:01 | 07:55 |
| GA-FG | * | * | * | * | * | * |
| GA-FS | 40.2 | 50.5 | * | * | 09:16 | 08:36 |
| HI-BP | 34.9 | 47.5 | 39.5 | 53.9 | 11:23 | 09:33 |
| HI-HI | 41.8 | 63.4 | 50.5 | 70.7 | 09:39 | 07:42 |
| ID | 54.4 | 68.7 | 33.0 | 45.1 | 10:33 | 08:14 |
| IL | 23.4 | 49.3 | * | * | 10:17 | 09:08 |
| IN | 29.8 | 47.5 | * | * | 17:30 | 08:10 |
| KY-FK | 29.6 | 37.9 | 22.0 | 32.8 | 12:14 | 10:40 |
| KY-HN | 32.7 | 55.5 | 37.3 | 53.5 | 10:58 | 08:36 |
| LA-CB | 31.8 | 51.3 | 30.3 | 41.0 | 09:09 | 08:08 |
| LA-CM | 29.6 | 36.1 | 25.3 | 42.9 | 09:03 | 07:27 |
| LA-GL | 26.4 | 37.9 | 20.7 | 29.4 | 09:17 | 10:52 |
| MD | 25.1 | 48.3 | 24.3 | 41.6 | 10:45 | 09:01 |
| MI | 41.1 | 53.8 | 32.1 | 56.7 | 08:27 | 07:48 |
| MS | 28.4 | 47.8 | 22.3 | 42.0 | 11:46 | 08:31 |
| MT | 35.7 | 49.9 | * | * | 10:35 | 08:35 |

**Table 2.14—Continued**

| Site | Curl-Ups | | Push-Ups | | 1-Mile Run | |
|------|---------|-------|---------|-------|-----------|-------|
|      | Initial | Final | Initial | Final | Initial   | Final |
| NC-NL | ^ | ^ | ^ | ^ | ^ | ^ |
| NC-S | 28.5 | 45.5 | 21.0 | 38.5 | 11:23 | 08:20 |
| NJ | 42.7 | 55.2 | 32.9 | 52.6 | 10:05 | 07:59 |
| NM | 31.7 | 50.9 | 34.3 | 64.1 | 08:04 | 06:18 |
| OK | 34.4 | 57.0 | 21.3 | 46.3 | 09:51 | 08:03 |
| OR | 37.6 | 55.4 | 18.4 | 29.2 | 12:08 | 12:06 |
| PR | 31.8 | 41.3 | 23.7 | 37.8 | 09:12 | 07:33 |
| SC | 38.6 | 44.5 | * | * | * | 08:02 |
| TX-E | ^ | ^ | ^ | ^ | ^ | ^ |
| TX-W | 34.5 | 47.0 | 24.3 | 51.8 | 12:20 | 10:38 |
| VA | 44.5 | * | 33.4 | * | 09:08 | * |
| WA | 46.4 | 73.2 | 17.2 | 44.9 | 10:45 | 07:39 |
| WI | 18.4 | 46.0 | 13.2 | 26.0 | 10:09 | 07:45 |
| WV | 33.0 | 53.7 | * | * | 10:04 | 07:08 |
| WY | 28.9 | 38.5 | 30.8 | 49.0 | 09:05 | 07:58 |

NOTE: Information in the table includes Class 44. Blanks in the table occur due to new sites (which have little historical data), or in cases where data were not reported or did not appear correct.

* Did not report.

^ Newly operational.

**Table 2.15**
**Residential Performance—Physical Fitness as Measured by the Average Number Completed and Time for Cadets per Site (Class 45)**

| Site | Curl-Ups | | Push-Ups | | 1-Mile Run | |
|------|---------|-------|---------|-------|-----------|-------|
|      | Initial | Final | Initial | Final | Initial   | Final |
| All Sites | 33.8 | 49.5 | 24.0 | 41.5 | 10:04 | 08:22 |
| AK | 32.5 | 47.3 | * | * | 10:53 | 07:51 |
| AR | 29.3 | 33.8 | 31.2 | 44.9 | 10:42 | 09:58 |
| CA-LA | 25.0 | 40.9 | 20.2 | 46.8 | 09:34 | 07:55 |
| CA-SL | 30.7 | 42.9 | 20.8 | 35.4 | 09:56 | 07:48 |
| D.C. | 26.1 | 30.9 | 19.4 | 25.1 | 13:34 | 11:44 |
| FL | 43.9 | 67.0 | 16.2 | 37.8 | 10:18 | 08:06 |
| GA-FG | 40.3 | 42.9 | 32.5 | 41.1 | 09:11 | 08:42 |
| GA-FS | 8.9 | 41.1 | * | * | 08:50 | 09:05 |
| HI-BP | 31.2 | 51.7 | 34.0 | 58.7 | 11:06 | 08:13 |
| HI-HI | 52.5 | 73.3 | 51.8 | 71.5 | 09:30 | 08:09 |
| ID | 49.8 | 70.1 | 25.1 | 44.7 | 10:12 | 07:51 |
| IL | 29.8 | 57.1 | 19.3 | 40.7 | 10:26 | 08:42 |
| IN | 30.3 | 52.8 | * | * | 10:19 | 08:16 |

**Table 2.15—Continued**

| Site | Curl-Ups | | Push-Ups | | 1-Mile Run | |
|---|---|---|---|---|---|---|
| | Initial | Final | Initial | Final | Initial | Final |
| KY-FK | 22.1 | 41.1 | 18.7 | 37.1 | 12:21 | 10:18 |
| KY-HN | 36.0 | 60.4 | 23.0 | 53.8 | 09:35 | 08:22 |
| LA-CB | 34.5 | 36.9 | 29.8 | 47.1 | 09:30 | 07:14 |
| LA-CM | * | * | * | * | * | * |
| LA-GL | 23.9 | 43.0 | 20.5 | 41.0 | 12:04 | 10:30 |
| MD | 31.5 | 58.3 | 25.5 | 46.9 | 12:09 | 08:39 |
| MI | 41.5 | 50.8 | 37.3 | 54.2 | 09:27 | 07:36 |
| MS | 35.3 | 53.2 | 21.8 | 46.3 | 10:51 | 07:55 |
| MT | 40.9 | 48.4 | * | * | 09:59 | 08:04 |
| NC-NL | 35.4 | 52.1 | 24.2 | 39.6 | 07:33 | 06:38 |
| NC-S | 30.4 | 42.5 | 21.9 | 36.0 | 10:27 | 07:50 |
| NJ | 36.4 | 44.9 | 29.1 | 45.7 | 07:32 | 06:46 |
| NM | 35.2 | 53.3 | 34.9 | 59.0 | 08:52 | 06:18 |
| OK | 38.4 | 45.2 | 23.3 | 36.5 | 09:59 | 08:56 |
| OR | 37.5 | 52.9 | 16.3 | 30.3 | 12:09 | 12:07 |
| PR | 35.7 | 44.3 | 27.5 | 39.0 | 08:29 | 07:28 |
| SC | 35.0 | 49.9 | * | * | 09:40 | 07:59 |
| TX-E | * | * | * | * | * | * |
| TX-W | 34.9 | 47.5 | * | * | 09:56 | 07:23 |
| VA | 36.9 | 50.7 | 25.7 | 53.1 | 09:09 | 07:37 |
| WA | 52.7 | 66.7 | 17.2 | 39.1 | 09:30 | 08:38 |
| WI | 18.2 | 42.9 | 12.1 | 21.0 | 09:42 | 08:08 |
| WV | 33.1 | 55.3 | * | * | 09:49 | 07:38 |
| WY | 28.0 | 34.8 | 26.1 | 33.1 | 09:15 | 08:24 |

SOURCE: RAND analyses based on data provided by ChalleNGe programs.

NOTE: Information in the table includes Class 45. Blanks in the table occur due to new sites (which have little historical data), or in cases where data were not reported or did not appear correct.

* Did not report.

and language areas. Sites also reported math scores independently. Math scores are lower than language scores (Math Battery scores are therefore somewhat lower than Total Battery scores), but the results are broadly similar if we use math scores in place of Total Battery scores. We also include a brief discussion of placements, as well as analysis of program costs and costs per cadet.

## Tests of Adult Basic Education Scores and Gain Scores

ChalleNGe graduates tend to make about two academic years of progress over the course of the 5.5-month Residential Phase. To characterize cadets' progress in more detail, we examined the extent to which cadets meet key benchmarks, defined as achieving at least ninth-grade (early high school) or eleventh-grade (late high school) levels of achievement. We focus on these metrics because these levels of achievement are linked to performance on the GED and AFQT tests.

Figure 2.1 characterizes the initial and final TABE scores among all ChalleNGe graduates from the two classes that began in 2015. Because there are very few cadets whose initial TABE scores are at the lowest grade levels, we combined levels into the categories presented in Table 2.1.[8] Figure 2.1 indicates that about 45 percent of cadets initially score at the elementary level and nearly 75 percent initially score at or below the middle school level. By graduation, just over half of cadets score at one of the high school levels and nearly one-third score at the eleventh-grade level or higher. Cadets therefore make considerable academic progress during ChalleNGe, and many cadets achieve key milestones while attending the program.

As already noted, past reports included only average TABE gains (by site) and that metric is problematic from a measurement perspective (see earlier TABE subsection). But the previous metric also fails to indicate the number or proportion of cadets who achieve key benchmarks. In contrast, the information presented in Figure 2.1 clearly indicates that about half of cadets score at the high school level (at or above grade 9) by the end of the program.[9]

**Figure 2.1**
**Initial and Final TABE Scores by Grade Level, ChalleNGe Graduates, Classes 44 and 45**

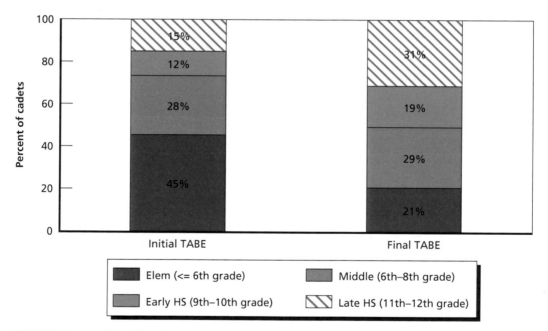

SOURCE: RAND analyses based on data provided by ChalleNGe sites for graduates from 2015 (Classes 44 and 45).
RAND RR1848-2.1

---

[8]  We recognize that the elementary grade levels include wide variation in terms of academic placement; in future analyses, we will examine movement between the elementary categories listed in Table 2.1.

[9]  Figure 2.1 includes only graduates; this is consistent with the information included in past reports. Roughly one-quarter of cadets do not complete the program (based on data provided by ChalleNGe sites for classes from 2015). However, our analyses indicate that initial TABE scores are not closely linked to the probability of graduation; cadets who begin the program at all academic levels graduate at similar rates. This suggests that the program is equally effective for a variety of participants, including those whose initial academic scores are quite low. There is no existing information on the relation-

Figure 2.1 certainly indicates academic progress, but from this figure it is not clear *which* cadets have the largest gains. When we examine cadets from each level separately, we find those who begin at the lowest grade levels make the most progress; indeed, cadets whose initial TABE scores are below the third-grade level make about three years of academic progress over the course of the program.

Figure 2.2 shows cadets' progress in more detail, indicating the category of initial and final TABE scores (colors indicate final TABE scores and match the colors used in Figure 2.1). Although cadets whose TABE Battery scores are initially at the elementary level make relatively large gains, Figure 2.2 indicates that these cadets are unlikely to score at the high school level by the end of the program. In contrast, the majority of those who begin at the middle school level reach one of the high school levels, and most who begin at the early high school level achieve a final score at the late high school level. This suggests that there is a trade-off inherent in admitting cadets with differing scores—gains are likely to be largest among cadets who have relatively low initial scores, but admitting cadets at the middle school level or higher means that more cadets will achieve scores in the high school levels. Recall that high school levels are associated with high probabilities of passing the AFQT as well as the GED (and, presumably, the HiSET). This also suggests that comparing gain test scores across sites without

**Figure 2.2**
**Progress by TABE Score Grade Level, ChalleNGe Graduates, Classes 44 and 45**

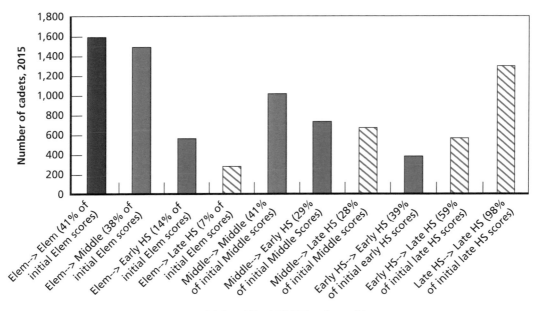

SOURCE: RAND analyses based on data provided by ChalleNGe sites for graduates from 2015 (Classes 44 and 45).
RAND *RR1848-2.2*

ship between initial TABE scores and longer-term outcomes (such as earnings). This is a potentially fruitful area for future analyses.

assessing the cadets' initial scores could be misleading because sites with lower-scoring cadets are likely to show higher gains.

Many factors influence test scores and test score gains. Examples include gender, age at which students leave school, peer effects, teacher characteristics, curriculum and school quality differences, regulations pertaining to education, and of course family background and resources.[10] While we do not have direct measures of many of these factors, we do have information on cadets' gender and on the age at which they entered ChalleNGe.[11] Next, we explore differences by these factors.

Figure 2.3 includes pre- and post-TABE scores, by level, for males and females. About three-quarters of ChalleNGe cadets, and about the same proportion of graduates, are male.[12] In the case of ChalleNGe graduates, the differences in TABE scores by gender are very small. Female cadets have slightly lower initial TABE scores; in particular, they are less likely to initially score at the late high school level.[13] But overall initial scores are roughly comparable; by the end of ChalleNGe, scores are nearly identical.

**Figure 2.3**
**Distribution of Initial and Final TABE Scores by Gender Among ChalleNGe Graduates, Classes 44 and 45**

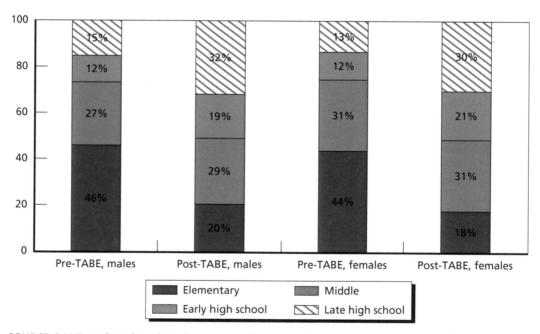

SOURCE: RAND analyses based on data provided by ChalleNGe sites for graduates from 2015 (Classes 44 and 45).
RAND RR1848-2.3

---

[10] See, among many other papers, Dee (2005).

[11] Note that cadets may have left a traditional high school right before entering ChalleNGe, or they may have entered ChalleNGe after some period of time.

[12] Because there are many more male graduates than female graduates, in this figure we show proportions for ease of comparison.

[13] While the difference is only about 2 percentage points, it is statistically significant at the 5-percent level, indicating that it is unlikely to have occurred by chance.

Figure 2.4 again presents information on initial and final TABE scores, but in this case we divide the cadets into groups based on their ages at the beginning of ChalleNGe. Slightly more than half of all ChalleNGe cadets, and about the same proportion of ChalleNGe graduates, are 16 when they enter the program; only about 11 percent enter at age 18. In terms of TABE scores, there are again only small differences (due to the differences in overall numbers by age, we show proportions). The initial and final scores of 18-year-olds are slightly lower than the scores of cadets who enter ChalleNGe at younger ages; in particular, cadets who enter at 18 are less likely to score at the late high school level initially or at the end of the Residential Phase of ChalleNGe.[14] However, the scores are generally comparable. Next we look at another key metric—placement within six months of completing ChalleNGe.

### Achieving Placement Within Six Months of Graduation

Within the ChalleNGe program, placement is considered a key metric. ChalleNGe staff work to keep in contact with graduates and their mentors both to assist the graduates in finding opportunities and to record the graduates' activities throughout the Post-Residential Phase. Placements may include military service, additional education, or working (as well as combinations of these, such as attending school and working). In this analysis, we include only graduates. As shown in Figure 2.5, the overall placement rate is 72 percent. Half of graduates who report having a placement six months after graduation are obtaining additional educa-

**Figure 2.4**
**Distribution of Initial and Final TABE Scores by Age at Entry Among ChalleNGe Graduates, Classes 44 and 45**

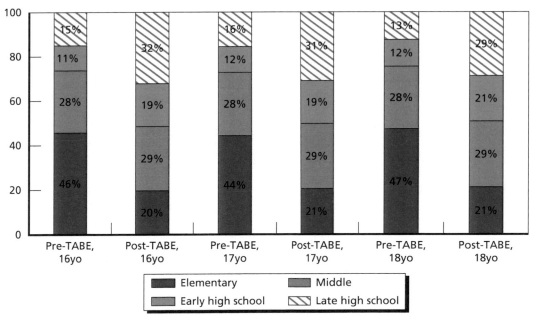

SOURCE: RAND analyses based on data provided by ChalleNGe sites for graduates from 2015 (Classes 44 and 45).
RAND RR1848-2.4

---

[14] The differences—2 percentage points at the beginning of ChalleNGe and 3 percentage points at the end—are statistically significant at the 5-percent level and therefore are unlikely to have occurred by chance.

**Figure 2.5**
**Placements Six Months After ChalleNGe Among Graduates,**
**Classes 44 and 45**

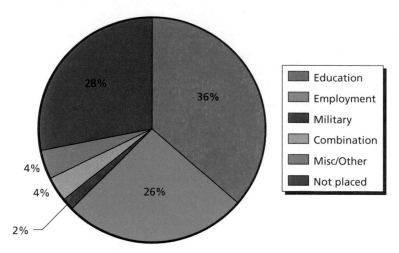

SOURCE: RAND analyses based on data provided by ChalleNGe sites.
This figure includes graduates from 2015 (Classes 44 and 45).
No data were provided for the Indiana program.
**RAND** *RR1848-2.5*

tion; many of the rest are employed, with smaller numbers serving in the military or reporting a combination of placements or some other sort of placement. We used simple regression models to test for a relationship between initial (or final) TABE score level and placement. While the regressions suggest that those who enter the ChalleNGe program with higher test scores also have a higher probability of placement, the relationship does not achieve statistical significance and thus may be due to chance. However, the relationship may be evident only in data collected over a longer period of time; we intend to further explore this relationship in future work.

We next provide potential benchmarks for ChalleNGe placement rates. The closest comparison to a six-month placement for a similar age group comes from a survey of recently graduated high school students participating in the USDOE's High School Longitudinal Study of 2009 (HSLS:09; NCES, 2016).[15] According to these data, in the fall after their final year of high school, approximately 75 percent of high school graduates are enrolled in postsecondary education, working, or doing both. Among GED holders, around 59 percent are enrolled in postsecondary education, working, or both. GED holders, however, are twice as likely to be working as are high school graduates; high school graduates are more likely to be continuing in postsecondary education. Dropouts are far less likely than either of these groups to be enrolled in any type of postsecondary education, but they are much more likely to be either working (36 percent) or to be neither taking classes nor working (20 percent). In general, these findings are comparable to graduates of the ChalleNGe program; indeed, the ChalleNGe placement

---

[15] The HSLS is a longitudinal study that began with a baseline survey of a nationally representative sample of ninth graders in the fall of 2009 and tracked them through high school with a follow-up in spring 2012 (most would have been juniors) and then an update (rather than a full follow-up) in November 2013 when most would have just completed high school. A second follow-up occurred in 2016 (Dalton et al., 2016).

rate of 72 percent resembles the activities of high school diploma graduates and exceeds the placement rate of GED holders in the HSLS.[16]

For further reference, the unemployment rate of 16–19-year-olds in late 2016 was about 15 percent; among those who have not completed high school, somewhat older data suggests an unemployment rate of about 30 percent.[17] Finally, young people who are discouraged by job market conditions and stop searching for work do not meet the official definition of unemployed and therefore would be considered out of the labor force. While ChalleNGe graduates appear to compare favorably to the groups included in HSLS, more detailed information on placements, and on desired placements among those who are not placed, would be helpful in determining more precise benchmarks.

## ChalleNGe Program Costs

States must pay at least 25 percent of the cost of ChalleNGe; as much as 75 percent is provided by the Department of Defense. Most sites follow the 25/75 split exactly, although a few states make larger contributions and a few sites receive funds from other sources (such as nonprofit foundations).[18]

ChalleNGe costs may vary for a number of reasons. For example, awarding high school diplomas may be more expensive than awarding other credentials. Older, more established sites may have lower (or higher) costs than newer sites. But site size is likely to be a driving factor. While sites with more cadets will be more costly, there are basic ("fixed") costs associated with the ChalleNGe program. Even the smallest sites must pay for administrative staff, teachers, facilities charges, and enough cadre to work with cadets 24 hours a day, 7 days a week. Expanding a program to include an additional platoon generally is associated with a sharp increase in total costs.[19] Recognizing that site cost is likely to vary with size, we analyze the cost data provided by the site by calculating a per-graduate cost. Figure 2.6 shows the per-graduate cost of each site. The average cost per graduate is roughly $20,000.[20]

---

[16] The HSLS study suggests that about 4 percent of recent high school graduates have joined the military (although this figure also includes those who are attending college and taking part in Reserve Officer Training Corps [ROTC] programs). Among nongraduates and GED holders, the proportion in the military is much lower, roughly 2 percent. These figures suggest that ChalleNGe graduates enlist at a slightly higher rate than other comparable young people, and at a rate that is at least as high as the enlistment rate among high school graduates.

[17] For the unemployment rate of all 16–19-year-olds, see "Economic News Release" (2017). The unemployment rate of 17–19-year-olds with no high school diploma was calculated from the American Community Survey 2014 five-year dataset using the person weights provided; for more information, see "American Community Survey (ACS)" (2017).

[18] Note that our analyses of ChalleNGe program costs used data reported by the sites. We requested the total amount of funding received from federal, state, and other sources. Some programs also receive various types of gifts or discounts from different sources. Examples include equipment transfers from the National Guard, discounts from various organizations, gifts from the local nonprofit arm of the program, or deferral of some of the costs of staff (e.g., through a local school district). Valuing such items is not straightforward; to the extent that programs vary in the manner in which they value such items, reported costs will differ. Indeed, we suspect that such additional resources may cause reported costs to be lower than actual total costs at some programs. In future data collection efforts, we will obtain detailed information about these resources. However, from the perspective of the Department of Defense, the cost data are likely quite accurate and the overall differences are likely to be small when compared to the total program budget.

[19] In future analyses, we will explore the staffing model and its relationship to costs in more detail.

[20] The average *site* has a slightly higher cost—roughly $23,000. This difference occurs because smaller sites have somewhat higher costs than sites with more graduates.

**Figure 2.6**
**Per-Graduate ChalleNGe Costs, 2015**

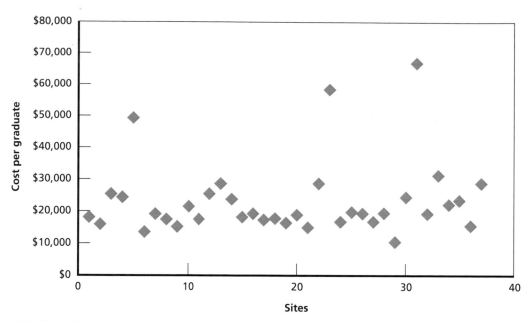

SOURCE: RAND analyses based on data provided by ChalleNGe sites.
NOTE: This figure includes graduates from 2015 (Classes 44 and 45).
RAND *RR1848-2.6*

Figure 2.6 indicates that per-graduate costs do vary across sites—but also that most sites have roughly similar costs while a few have vastly different ones. Next, we use the data provided by the sites to model cost per graduate. We want to separate costs related to site size from those related to other factors (site age, type of credential awarded). To do this, we use a regression model—which, with variation across all the factors of interest, we can use to separate effects related to size from those related to age or credential.[21] When we do so, we find that size is the driving factor behind cost per graduate—and that sites with fewer than 150 graduates per year cost substantially more than larger sites. While newer sites and those that award high school diplomas have higher costs on average, the differences become small and insignificant in our regression model. This indicates that size is the driving factor in costs.

Figure 2.7 shows the relationship between the number of graduates per year and the per-graduate cost of a ChalleNGe site. The data indicate that cost per graduate generally decreases with site size, but that the largest drop in cost occurs between 50 and 150 graduates per year (all sites had at least 50 graduates per year). Beyond 150 graduates per year, costs are fairly constant.[22] These results suggest that encouraging sites to target at least 150 or more graduates per

---

[21] We experimented with several specifications that included type of credential and region, but the results indicated that only the number of graduates was statistically significantly related to cost. The results we report here are from a regression including the number of graduates, as well as squared and cubed measures of the number of graduates.

[22] Note that these results include only two classes of data; we plan to repeat this analysis on updated data in future reports. Also, note that cost per graduate is a function of the number of graduates, and of the total budgeted for the program. An important factor in determining the total budget is the total number of *planned* graduates (the target or benchmark). Programs with fewer than the planned number of graduates will, of course, have higher-than-average costs per graduate, and these programs will generally be quite small. Future analyses will explore the relationship between the planned (or

**Figure 2.7**
**Relationship Between Per-Graduate Costs and Number of Graduates, 2015**

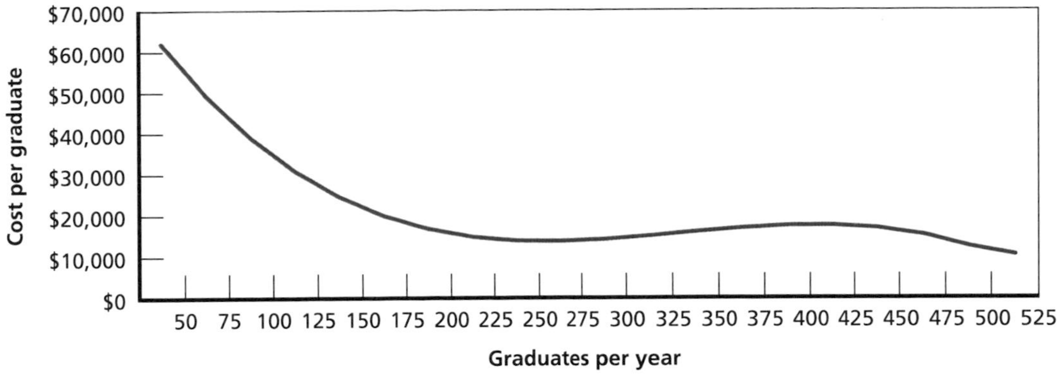

SOURCE: RAND analyses based on data provided by ChalleNGe sites.
NOTE: This figure includes graduates from Classes 44 and 45. Data are regression-adjusted, including controls of number of graduates as well as squared and cubic terms.
**RAND** *RR1848-2.7*

year has the potential to increase cost-effectiveness across the ChalleNGe program. Of course, some new sites may struggle initially to achieve this benchmark. However, the cost figures from 2015 suggest that encouraging sites to achieve this benchmark is an appropriate policy to control costs.

We do note that most sites reported at least 150 graduates per year in 2015; only six sites reported fewer graduates. Also, the sites with the highest costs are quite small; all three of those with the highest costs had fewer than 100 graduates per year. As such, while these sites are disproportionately costly, they still represent only 6 percent of the total ChalleNGe program costs.

## How Do ChalleNGe Costs Compare with Those of Similar Programs?

There are no obvious benchmarks for ChalleNGe costs. Job Corps is a program that is similar to ChalleNGe in some ways, but some participants spend substantially more than 5.5 months at Job Corps (which is self-paced). In 2001, the cost per *participant* for that program was about $17,000 (McConnell and Glazerman, 2001). This suggests that the cost of Job Corps was about $23,000 in 2015, and that Job Corps is more expensive than ChalleNGe; it is not clear how much of the difference is based on the self-paced aspect of Job Corps or the longer time period involved.[23]

Public schools offer another potential comparison point. Although students spend only limited hours in school, they attend for most of the year. The average cost per pupil in 2015–2016 was about $13,000.[24] For a final reference point, the cost of incarceration is roughly

---

benchmarked) number of graduates and other relevant factors to determine which programs are likely to fall short of this benchmark.

[23] The $23,000 figure assumes costs have increased proportional to inflation. The inflation adjustment is based on the Consumer Price Index for All Urban Consumers (CPI-U); see "Consumer Price Index" (n.d.).

[24] The expenditure per pupil in average daily attendance was $13,373; total expenditure based on fall enrollment was $12,509 (see National Center for Education Statistics, 2016). Constant dollars based on monthly CPI-U, adjusted to reflect the period of time included in the school year.

$31,000 per year.[25] Recall also that ChalleNGe has been found to be effective in the sense that participants gain more education and have higher earnings than similar youth who do not participate. Indeed, the gains are substantial and indicate that each dollar invested in ChalleNGe results in over $2.60 in returns, mostly due to participants' increased future earnings (Millenky et al., 2011; Perez-Arce et al., 2012).

While the benchmarks that exist suggest that the cost of ChalleNGe is within the range of somewhat similar programs and that the program has been found to be cost-effective, the cost data indicate that some sites have higher-than-average costs, and that program size is the driver behind this trend. Therefore, increasing the size of the smallest sites has the potential to improve the cost-effectiveness (and lower the average costs) of ChalleNGe.

## Conclusion

In this chapter, we presented information collected from individual ChalleNGe sites to provide a snapshot of the program in 2015 and to support the program's report to Congress. We also presented information that uses existing data on TABE scores in a new way—to determine the number of cadets who achieved key levels of academic achievement while attending ChalleNGe. Cadets whose TABE scores placed them in the (early or late) high school range have substantially higher chances than others of passing the GED exam. We find that most cadets who enter ChalleNGe at the middle school level or higher achieve these TABE scores. This suggests that programs may wish to provide additional support to cadets whose initial TABE scores are at or below the middle school level. While our average TABE gains are similar to those noted in earlier reports, the more detailed metrics provide additional information to programs and decisionmakers. The data indicate that cadets make considerable progress in a number of areas while attending ChalleNGe. The ChalleNGe placement rate of 72 percent resembles that of high school diploma graduates and exceeds the placement rate of GED holders in the HSLS. This is similar to placement rates reported from earlier classes.

Finally, we included some analysis of the cost data. Our analysis of cost data provided by the sites indicates that, while most ChalleNGe sites have somewhat similar average costs per graduate, a few sites have costs that are much higher. We explored several possible reasons for cost variation and found that, rather than differences in sites' ages or credentials awarded, size (number of graduates) is the driver of cost. ChalleNGe sites with fewer than 150 graduates per year have substantially higher costs than other sites. As a first step to reducing average costs, programs should consider options to expand the number of cadets served.

The program data demonstrate variation across many measures—number of applicants, completion rate, staff-to-cadet ratio, among others. In states such as Georgia, Louisiana, Oklahoma, and Illinois, there were over 350 applicants across both classes, and less than 150 applicants in Idaho, Hawaii (Hilo), and Montana. By way of comparison, on average across sites and across classes, around 250 applicants applied. Completion rates vary both by site and

---

[25] This figure varies from a low of $15,000 in states like Indiana and Kentucky to a high of $60,000 in New York. Among the factors that may account for differential costs between states is the extent of overcrowding of correctional facilities, or conversely, reducing the inmate population while maintaining the same level of operating costs. Some states have high incarceration rates of low-level offenders and rely on local jails to keep state-sentenced inmates where programming is more limited. For a careful calculation of costs of incarceration, see Henrichson and Delaney (2012).

class. For example, completion rates of 60 percent or less for Class 44 are reported by D.C. (56 percent), Georgia—Fort Gordon (60 percent), Georgia—Fort Stewart (60 percent), and Virginia (57 percent). In all these cases except for one, these sites reported higher completion rates for Class 45: D.C. (73 percent), Georgia—Fort Gordon (60 percent), Georgia—Fort Stewart (68 percent), and Virginia (64 percent). Other sites reporting completion rates of 60 percent or less for Class 45, which were lower than completion rates for Class 44, include Kentucky—Fort Knox (52 percent) versus 74 percent for Class 44 and Oklahoma (55 percent) versus 63 percent for Class 44. Some sites, such as Florida, Maryland, and Puerto Rico, report 100 percent completion in both Classes 44 and 45. The number of cadets per staff member also varies—as low as two cadets for every staff member to as high as seven cadets for every staff member.[26] A key part of future analyses will include delving into these differences to determine the root causes; such findings could provide guidance on how best to serve future cadets and use resources effectively.

In the next chapter, we shift the focus from addressing the first purpose of our report—providing a snapshot of recent ChalleNGe performance—to the second purpose—to begin the process of developing a richer set of metrics that will describe the long-term effects ChalleNGe has on participants after they leave the program. The framework presented in the next chapter will provide the starting point for subsequent annual reports.

---

[26] The cadet-to-staff ratio only includes instructors and cadre (both full-time and part-time) in the computation since they most frequently interact with cadets. Additional staff members not included in the computation are those classified as administrative staff and other staff.

# CHAPTER THREE

# Initial Framework for Measuring the Longer-Term Outcomes

This chapter presents our framework for measuring longer-term outcomes on the ChalleNGe program moving forward. The chapter is divided into two parts.

The first part presents the TOC, which describes the elements of the program and the underlying mechanism that will lead to the positive change in young people's lives. The TOC is followed by a ChalleNGe program logic model, which illustrates how program resources, activities, services, and outputs are linked to short-, medium-, and long-term outcomes (Gonzalez et al., 2016). The TOC expresses the mechanisms through which ChalleNGe is designed to work. Along with providing an explanation for how the ChalleNGe program works, the TOC is a first step in building a program logic model (Knowlton and Phillips, 2009). The program logic model allows us to categorize the resources and activities of the program and to divide program effects into those that would be expected to occur immediately versus those expected to occur at future points in time. This tool provides a detailed framework for developing metrics (Gonzalez et al., 2016). We also discuss the implications of the TOC and the logic model for ChalleNGe measurement and evaluation.

The second part of the chapter focuses on data issues. Because additional data collection will be needed to link program attributes to longer-term outcomes, we provide a brief overview of some of the relevant barriers to and strategies for collecting these data. Future reports will develop specific data collection strategies, with specific examples drawn from relevant existing sites and notes on data tracking issues that are likely to be especially pertinent to ChalleNGe.

## Program Models

### ChalleNGe Program Theory of Change Model

Advocates for providing youth mentoring opportunities argue that young people today face both tremendous opportunities and tremendous pressures, and many appear not to receive the type of guidance they need to make informed choices (Bruce and Bridgeland, 2014). Some of the most basic examples of poor choices that young people make are reflected in risky behaviors, such as not wearing a seatbelt (10 percent), riding in a car operated by someone who had been drinking alcohol (28 percent), carrying a weapon (18 percent), and using alcohol or marijuana (42 and 21 percent, respectively).[1] These behaviors place young people at risk of potential run-ins with the law, further jeopardizing their future well-being—not to

---

[1] Behaviors reported among those aged 10–24 within the previous 30 days. It is alarming to note that 6 percent had attempted suicide within the previous 12 months (Eaton et al., 2010).

mention that these behaviors can be life-threatening (Freeman and Simonsen, 2015; Amin et al., 2016). Young people also face critical decisions about their education, training, and career choices that will have a direct link to their long-term livelihood, financial independence, and successful transition into adulthood. Socioeconomically disadvantaged young people are particularly at risk for not sufficiently investing in developing their skills, including completing high school so that they can continue on to postsecondary education and training—an increasingly necessary step before being deemed "job ready" (Freeman and Simonsen, 2015).

Programs that target just one aspect of these problems ignore the interconnectedness of many of the pressures that young people face. For example, programs targeting only the cognitive skills necessary to do a job (i.e., "hard" skills) while ignoring other essential socioemotional or noncognitive (also referred to as "soft") skills may not sufficiently prepare young people both to secure and keep jobs (Heckman, 2000). Moreover, many of these programs fail to take advantage of the positive encouragement that peers and mentors can provide (Hossain and Bloom, 2015).

In order to better understand the ChalleNGe intervention, we developed a TOC—a useful tool for understanding a complex social problem and conceptualizing the mechanisms through which solutions can be developed to address the problem. The TOC for the ChalleNGe program is based on the premise that giving young people a second chance through an intensive, residential-based, regimented program—without which young people would face too many distractions—and providing a scaffolding of ongoing mentorship after the program has been completed will increase the likelihood that program participants can achieve success in work and life.

To ensure that the program is achieving its goals, the TOC also provides a foundation for identifying the types of outcomes that will be measured to track progress. The TOC is a first step to developing a monitoring and evaluation (M&E) system, including metrics or indicators that will be measured to ensure implementation fidelity to the design (the "monitoring" component of M&E) and to assess whether the program is having the desired effect on the outcomes (the "evaluation" component of M&E).

The ChalleNGe program is built on the whole-person concept, with a focus on developing the cognitive, emotional, and physical aspects of a young person in order to set him or her on a more productive life course (Price, 2010). The eight core components of leadership/followership, service to the community, job skills, academic excellence, responsible citizenship, life-coping skills, health and hygiene, and physical fitness have considerable overlap and can be broadly grouped into five tenets that contribute toward instilling positive, prosocial habits that help a young person achieve a rewarding, fulfilling life:

- Develop leadership or followership behaviors through discipline, hard work, and persistence.
- Engage in activities that promote good physical health.
- Act as a responsible citizen and build strong linkages to the community through service and participation.
- Attain academic skills and credentials to create job-readiness and the potential for success in the labor market.
- Strengthen socioemotional skills to build life-coping strategies.

We developed these groupings by combining core components that could be measured jointly; for example, eventual earnings could be used as one metric of academic excellence and job skills.[2] Mentoring does not appear as a tenet because it does not constitute a single skill that cadets are expected to obtain; rather, mentoring is intended to support the graduates as they apply newly acquired skills and competencies through the postresidential transition. Figure 3.1 summarizes the TOC on which the ChalleNGe program is based.

The TOC is best explained by illustrating the phases of the ChalleNGe program in Figure 3.1. During the Residential Phase (dark blue box on the far left), the five tenets are shown as the central elements of the training and development programs that are undertaken as part of the program. The program seeks to develop positive, prosocial habits and attributes in young people by giving them a transformative life experience. It takes young people out of distracting or negative contexts and provides time to develop both the physical and mental skills needed to shield them from deleterious outside influences. Cadets are expected to actively participate in community service activities, complete academic requirements, develop life-coping skills, learn how to lead and/or follow, and replace unhealthy behaviors (including drug and alcohol use) with healthy behaviors (exercise, healthy diets). Cadets are placed in closely monitored settings where they can develop healthy relationships with other individuals, many of whom share the same background and life experiences as they do and hold similar long-term aspirations for achievement and life success. Once these young people are exposed to these new experiences, they are more likely to develop new skills, attitudes, and a positive outlook on life.

Even after cadets complete the Residential Phase and graduate from the program, they may continue to face choices that can set them back on the wrong path in life. The TOC posits that a key to reinforcing and sustaining this success is matching these young people to adults with life experience who can help them navigate difficult life choices and smooth the transition into adult roles. Both the skills- and character-building efforts buttressed by the long-term mentorship are intended to propel a young person toward experiencing a rewarding, productive life. The Post-Residential Phase is represented by the blue box in the middle of the diagram where the mentor-cadet relationship reinforces cadets' community engagement and investment. It helps them to stay on track to achieve college and career readiness; develop character traits that help them cope with life's challenges; build personal and professional networking relationships; and engage in healthy living. In turn, this is expected to serve them beyond the influence of ChalleNGe to achieve the elements of a rewarding and productive life, including a feeling of belonging to the community, achieving economic self-sufficiency, developing healthy personal relationships and supportive professional networks, and achieving mental and physical well-being. One clear implication from the TOC is that the ChalleNGe program appears to work by focusing on many aspects of a young person's life and behavior. This suggests that metrics should span the eight core components (or the five tenets shown in the TOC).

## ChalleNGe Program Logic Model

We next developed a program logic model, which delineates the inputs, processes or activities, expected outputs, and desired outcomes of a program (Shakman and Rodriguez, 2015).

---

[2] We formed these tenets to summarize the eight core components of ChalleNGe: academic excellence, career explorations (job skills), health and hygiene, leadership/followership, life coping skills, physical fitness, service to community, and responsible citizenship (Price, 2010; "Core Components," n.d.).

**Figure 3.1**
**National Guard Youth ChalleNGe Program Theory of Change Model**

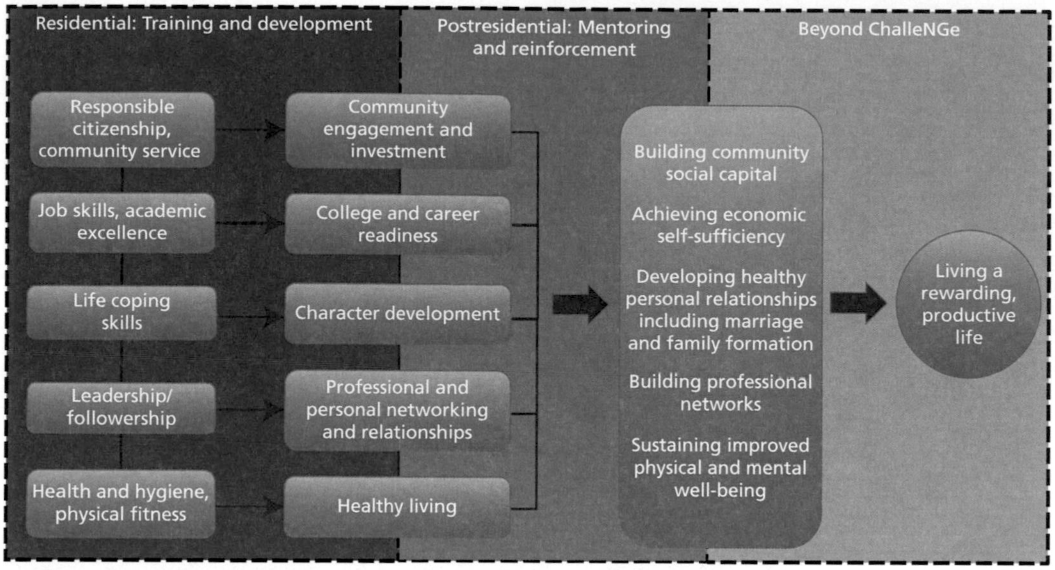

SOURCE: RAND analyses based partly on information provided by
"National Guard Youth ChalleNGe," (n.d.).
RAND *RR1848-3.1*

The program logic model, while based on many of the same ideas as the TOC, includes more information in terms of the program's inputs and outputs and lays out expected results in more detail. The program logic model also emphasizes the temporal aspects of ChalleNGe and its influence on participants.

Program logic models are a useful way of specifying the reasoning behind program structure and activities and how those activities are connected to expected program results (Knowlton and Phillips, 2009). They are used to illustrate how program resources, activities, services (inputs), and direct products of services (outputs) are designed to produce short-term, medium-term, and long-term outcomes. These models also identify broader community impacts that should result from program activities and services (Knowlton and Phillips, 2009). As such, they serve to communicate how a program contributes not only to the specific needs and outcomes of program participants, but also to the broader community and society at large. Program logic models also serve as a blueprint for evaluating how effectively a program is meeting its expected goals.

Figure 3.2 displays the program logic model we developed for the ChalleNGe program. This logic model was informed by a review of program documentation and annual reports, followed by site visits to two ChalleNGe locations (the Mountaineer ChalleNGe Academy in West Virginia and the Gillis Long ChalleNGe site in Louisiana).

Program inputs (the resources needed to administer the program) include policy and planning materials to guide program activities and the assets needed to house and instruct cadets. Program activities include Acclimation Period orientation activities, undertaken to prepare cadets for ChalleNGe (e.g., performing physical exams, instructing cadets on program standards and expectations). The Acclimation Period activities feed directly into program activities during the Residential Phase. Program outputs include those related to cadet

**Figure 3.2**
**Program Logic Model Describing the National Guard Youth ChalleNGe Program**

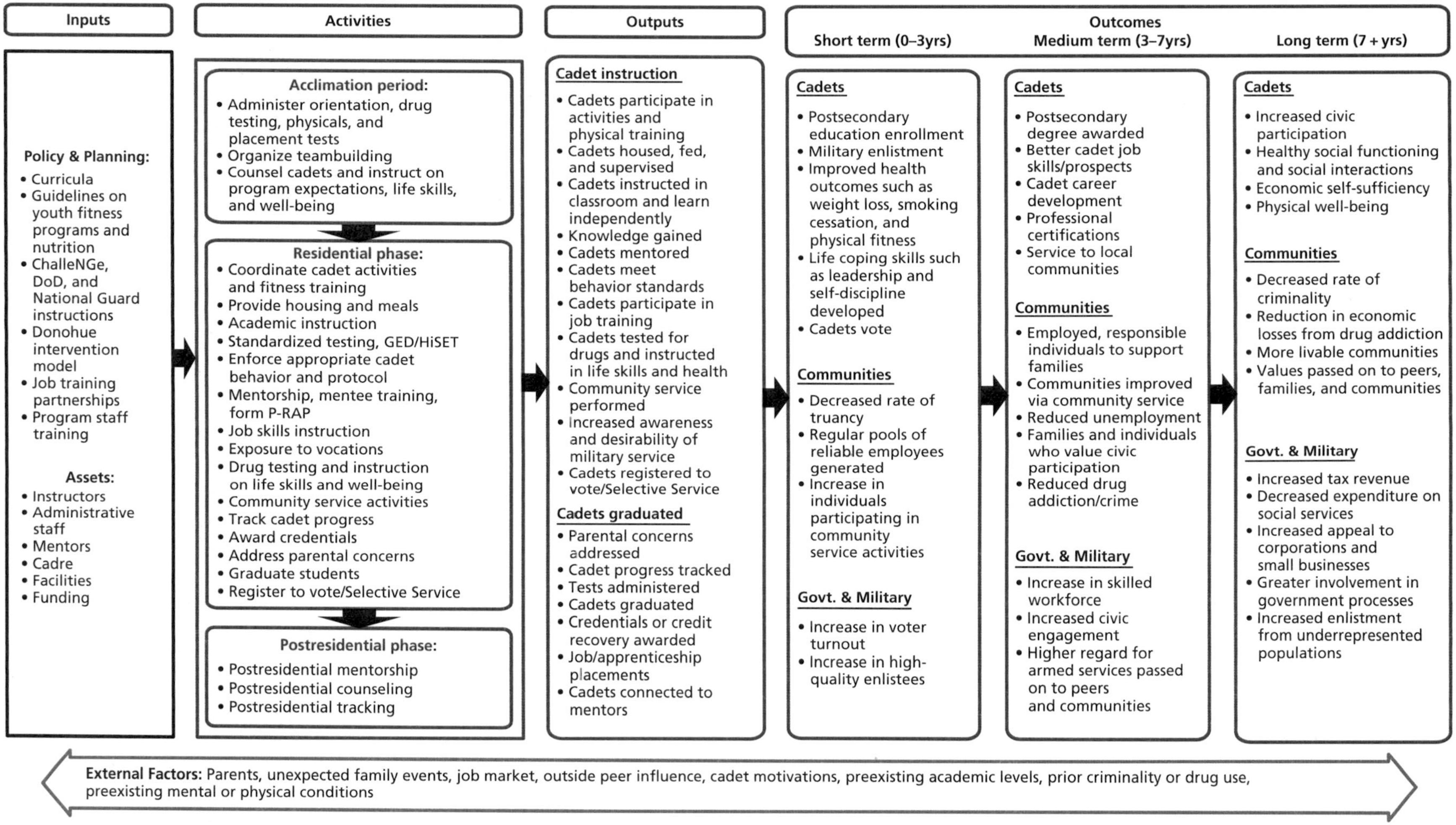

| Inputs | Activities | Outputs | Outcomes | | |
|---|---|---|---|---|---|
| | | | Short term (0–3yrs) | Medium term (3–7yrs) | Long term (7 + yrs) |

**Inputs**

**Policy & Planning:**
- Curricula
- Guidelines on youth fitness programs and nutrition
- ChalleNGe, DoD, and National Guard instructions
- Donohue intervention model
- Job training partnerships
- Program staff training

**Assets:**
- Instructors
- Administrative staff
- Mentors
- Cadre
- Facilities
- Funding

**Activities**

**Acclimation period:**
- Administer orientation, drug testing, physicals, and placement tests
- Organize teambuilding
- Counsel cadets and instruct on program expectations, life skills, and well-being

**Residential phase:**
- Coordinate cadet activities and fitness training
- Provide housing and meals
- Academic instruction
- Standardized testing, GED/HiSET
- Enforce appropriate cadet behavior and protocol
- Mentorship, mentee training, form P-RAP
- Job skills instruction
- Exposure to vocations
- Drug testing and instruction on life skills and well-being
- Community service activities
- Track cadet progress
- Award credentials
- Address parental concerns
- Graduate students
- Register to vote/Selective Service

**Postresidential phase:**
- Postresidential mentorship
- Postresidential counseling
- Postresidential tracking

**Outputs**

**Cadet instruction**
- Cadets participate in activities and physical training
- Cadets housed, fed, and supervised
- Cadets instructed in classroom and learn independently
- Knowledge gained
- Cadets mentored
- Cadets meet behavior standards
- Cadets participate in job training
- Cadets tested for drugs and instructed in life skills and health
- Community service performed
- Increased awareness and desirability of military service
- Cadets registered to vote/Selective Service

**Cadets graduated**
- Parental concerns addressed
- Cadet progress tracked
- Tests administered
- Cadets graduated
- Credentials or credit recovery achieved
- Job/apprenticeship placements
- Cadets connected to mentors

**Short term (0–3yrs)**

**Cadets**
- Postsecondary education enrollment
- Military enlistment
- Improved health outcomes such as weight loss, smoking cessation, and physical fitness
- Life coping skills such as leadership and self-discipline developed
- Cadets vote

**Communities**
- Decreased rate of truancy
- Regular pools of reliable employees generated
- Increase in individuals participating in community service activities

**Govt. & Military**
- Increase in voter turnout
- Increase in high-quality enlistees

**Medium term (3–7yrs)**

**Cadets**
- Postsecondary degree awarded
- Better cadet job skills/prospects
- Cadet career development
- Professional certifications
- Service to local communities

**Communities**
- Employed, responsible individuals to support families
- Communities improved via community service
- Reduced unemployment
- Families and individuals who value civic participation
- Reduced drug addiction/crime

**Govt. & Military**
- Increase in skilled workforce
- Increased civic engagement
- Higher regard for armed services passed on to peers and communities

**Long term (7 + yrs)**

**Cadets**
- Increased civic participation
- Healthy social functioning and social interactions
- Economic self-sufficiency
- Physical well-being

**Communities**
- Decreased rate of criminality
- Reduction in economic losses from drug addiction
- More livable communities
- Values passed on to peers, families, and communities

**Govt. & Military**
- Increased tax revenue
- Decreased expenditure on social services
- Increased appeal to corporations and small businesses
- Greater involvement in government processes
- Increased enlistment from underrepresented populations

**External Factors:** Parents, unexpected family events, job market, outside peer influence, cadet motivations, preexisting academic levels, prior criminality or drug use, preexisting mental or physical conditions

SOURCE: RAND analyses based on information collected from ChalleNGe sites.
NOTES: The Donohue intervention model was the initial design and description of the ChalleNGe program (Price, 2010). GED and HiSET credentials are awarded based on performance on standardized tests. The P-RAP is the Post-Residential Action Plan, designed to support planning and goal development among cadets.
RAND RR1848-3.2

instruction activities (e.g., housing, instructing, and mentoring cadets) and those related to the end process of graduating cadets (e.g., administering standardized tests, awarding credentials, placing cadets). Outcomes expected to result from program completion include those in the short term (within three years of graduation), medium term (within three to seven years of graduation), and long term (seven years or more after graduation). These include positive outcomes for the cadets themselves and their families (e.g., better job skills and job prospects), as well as for their communities, government, and the military (e.g., an increase in individuals participating in community service activities, greater tax revenue, increased military enlistment from underrepresented populations). Understanding the dynamic flow of the relationships between and among the inputs, outputs, and outcomes, and measuring the expected connections among these components will allow for systematic evaluations of the ChalleNGe program (W. K. Kellogg Foundation, 2006).

Logic models serve primarily as tools to assist us in developing new, improved metrics. But we also note that logic models can be useful tools to communicate key aspects of a program to a variety of stakeholders. We plan to present the program logic model to ChalleNGe directors and other program staff and collect their feedback. We will continue to refine and expand upon the current program logic model in future reports.

### Implications of the Logic Models

Recall that ChalleNGe's mission is to produce graduates who are successful, productive citizens in the years after they complete the program. The research that established the effectiveness of ChalleNGe on job performance and earnings, and the cost-benefit calculations associated with that research, focused on longer-term outcomes (see Chapter One). This suggests that while outcomes and short-term outputs are key aspects of a program's *performance*, determining the extent to which ChalleNGe is meeting its *mission* will require collecting longer-term outcomes. In contrast, the existing metrics presented in Chapter Two tend to focus on the left-hand side of the logic model—inputs, activities, and outputs. The TOC, meanwhile, suggests that ChalleNGe works by focusing on many aspects of the individual. Therefore, effective metrics are likely to include various aspects of the core components rather than focusing solely or mostly on, for example, academic achievement.

In summary, our TOC and program logic model both provide useful guidelines for developing improved metrics for the ChalleNGe program. Current metrics often focus on early aspects of the program (resources, inputs, activities). Specific examples of such metrics include the number of cadets admitted, graduates, credentials awarded, and hours of community service, as well as the decrease in one-mile-run times. We have collected and reported on many of these metrics (see Chapter Two). These metrics do include information from several aspects of the ChalleNGe program (such as physical fitness and community service), but comparing this list to the logic model suggests that future metrics should also include increased focus on longer-term outcomes and impacts, as these metrics are more closely related to the ChalleNGe program's mission. Especially in terms of academic progress, there is a significant emphasis on the short-term aspects of the program. This includes metrics of credentials awarded and TABE scores.[3] However, there is virtually no information collected on how these metrics relate to

---

[3]  Recall that past TABE scores were reported as average gains; this metric is problematic because it can inappropriately identify growth (Lindholm-Leary and Hargett, 2006). For this reason, we formulate new benchmarks from the TABE data. These benchmarks offer more information than previous metrics, but they still focus on short-term academic progress.

longer-term impacts (such as eventual educational attainment or earnings). In addition, future metrics could include more aspects of the eight core components.

## Data Collection: Barriers and Strategies

Linking ChalleNGe program attributes to longer-term outcomes will require different or additional data compared to those available today. Of course, data collection can be both expensive and time-consuming. We identified some barriers to additional data collection, along with strategies for addressing those barriers.

### Data Collection in Similar Programs

Numerous programs funded by the USDOE focus on ABE/ASE. In some ways, such programs appear to have little in common with ChalleNGe. They generally are not residential and are not limited to teens; also, these programs emphasize academic achievement rather than other aspects of ChalleNGe's core components. However, USDOE requires learning gains in these programs to be measured with the TABE or a similar approved assessment, and also requires a focus on four postprogram outcomes: high school or recognized equivalent completion; entrance into postsecondary education or training; entrance into the workforce; and retention of employment (NRS, 2015). These outcomes have a great deal of overlap with the types of information that ChalleNGe sites will need to collect to measure longer-term program outcomes. Therefore, strategies developed for these programs offer guidance for data collection strategies that could be used with ChalleNGe.

The latest annual report to Congress on the Adult Education and Family Literacy Act (AEFLA) of 1998 (USDOE, 2015) identifies specific methods for tracking post-ABE/ASE outcomes. According to this document, states measure these outcomes with two primary sources of information: surveys and administrative data. Programs administer follow-up surveys to students to collect self-reported information on the outcomes of interest.[4] The second, and more common, method of identifying post-ABE/ASE outcomes is through administrative data sources. These include the Statewide Longitudinal Data System (SLDS) from education departments and state unemployment insurance (UI) wage records. In some cases, states match self-response data to administrative data to identify postparticipation outcomes.

Tracking individuals through multiple data systems requires administrators of each system to collect compatible information to match records.[5] For ABE/ASE programs operating in public school districts or community colleges, these linkages are natural extensions of work already being done in many states through the implementation of SLDSs. Here, a student's educational experience is connected from early childhood education through postsecondary completion, including receipt of high school equivalency degrees and any ongoing educational enrollment or course taking. Building relationships with local school districts is identified as a key program effort that all programs/centers aimed at reengaging students into the education system and subsequent workforce should undergo at startup (Rennie-Hill et al., 2014). For programs operating outside the administrative boundaries of the traditional education

---

[4]   The NRS, which is the accountability body of ABE/ASE in the United States, has established guidelines for conducting these surveys (Division of Adult Education and Literacy Office of Vocational and Adult Education, 2015).

[5]   For a quick introduction to connecting data across multiple sources, see Institute of Education Sciences (2014).

system, additional steps must be taken to connect the appropriate information across systems. The most basic of these steps is to ensure there is a unique identifier established for every participant that can link the participant to the other state administrative data records, typically a Social Security Number or combination of date of birth, name, gender, and driver's license or state ID number. These unique identifiers allow for the sharing and connecting of critical data to measure program outcomes and support students. Should the ChalleNGe program follow this model, it would be necessary to develop agreements with other agencies (e.g., state departments of education, labor, corrections) to track information on graduates in this way. Such an approach is most likely impractical for ChalleNGe; the program faces several specific data tracking barriers, which we discuss next.

### Data Tracking Barriers Faced by ChalleNGe

The ChalleNGe program is likely to face three data-related barriers that are not common to state-run ABE/ASE programs and one barrier that likely affects all ABE/ASE programs in efforts to collect data.

- ChalleNGe is not connected with or conducted through state departments of education or labor and therefore may not have the ease of access to administrative data systems other ABE/ASE programs have.
- ChalleNGe sites serve individuals from multiple cities or counties within a state and numerous school districts, each with their own process for sharing data and personnel/wage records.
- ChalleNGe operates in multiple states, all of which have their own regulations for sharing data, collect information in ways that differ across contexts, and follow different state regulations regarding testing and outcomes required for ABE/ASE programs. The information being asked for and collected by ChalleNGe may therefore not be compatible across all locales.
- Individuals are mobile and may not remain in the state where they participated in ChalleNGe. Economic and educational opportunity, enlistment in the military, and life circumstances more broadly may require individuals to move to another state following program participation. Therefore, even when relationships are established with local government agencies, not all of the information needed to track participants will be available to ChalleNGe.

Due in part to these barriers, ChalleNGe currently tracks postprogram outcomes through a different methodology: postresidential counselors at each ChalleNGe program are responsible for maintaining contact with recent graduates for one year. Counselors communicate with graduates through phone calls and email, and they leverage mentors' relationships with program graduates to determine graduates' current activities. A primary emphasis of this tracking is to ensure that graduates maintain a *placement* (being enrolled in school, being employed, serving in the military, or some combination of these). Counselors, who also assist graduates in obtaining employment or enrolling in school, indicate that maintaining contact with graduates can be difficult because these young people frequently move, obtain new phone numbers, and change their placement status (e.g., by quitting a job or enrolling in a new school). Placement information collected by counselors is not detailed; for example, there is no information about wages or earnings, only employment status. Including additional detail in the placement information would allow measurement of more finely grained outcomes, such as the number of ChalleNGe graduates who are self-sustaining based on their earnings and the extent to which graduates tend to remain in a single job versus switching jobs (during past eras, job-switching among younger workers was associated with substantial wage growth; see, e.g., Topel and Ward, 1988).

Finally, we note that the National Student Clearinghouse (NSC) offers a somewhat different option for linking ChalleNGe data to information on longer-term outcomes. This data source exists for research purposes and includes information on students who enroll in postsecondary institutions. With identifying information, it would be possible to match a list of ChalleNGe cadets against the NSC database and determine how many cadets eventually enrolled in postsecondary institutions and completed degrees or credentials. While doing so would require information from cadets who completed the program a number of years ago (so they would have time to enroll and complete their study), this method offers a way to measure longer-term postsecondary outcomes on many cadets and would not require extensive agreements with other agencies.

## ChalleNGe-Specific Approaches to Data Collection

In the event that the ChalleNGe program cannot, or chooses not to, access administrative data from external agencies, ChalleNGe will need to rely on a combination of techniques. First, the role of counselors could be altered or expanded to collect more, or more detailed, information about cadets' placements and other longer-term outcomes. Second, surveys could prove helpful in extending the window of data collection.[6] In particular, online surveys may offer a cost-effective method to reach many participants and to ask a variety of detailed questions related to longer-term impact and outcomes. The NRS implementation guide suggests several relevant practices for improving survey response rates:

- Inform participants at program entry about follow-up survey and expected participation.
- Collect extensive contact information at program entry, including for family and friends likely to know participant's whereabouts over time (for example, after moving).
- Update contact information on a regular basis to ensure most current and accurate details.
- Request that participants initiate the updating of personal information when a move or change of contact information occurs. (NRS, n.d.)

RAND is currently investigating these practices in national programs that must collect similar follow-up information. Our findings in this area will be documented in future reports.

---

[6] The NRS (n.d.) has established relevant guidelines for conducting such surveys. On population-wide surveys, a response rate of 50 percent is the target established by the NRS in order for programs to draw representative conclusions about outcomes.

# CHAPTER FOUR

# Concluding Thoughts

The National Guard Youth ChalleNGe program continues to have a positive influence on the lives of thousands of young people each year and does so in a cost-effective manner. However, there is substantial variation across program sites on many metrics (Tables 2.3–2.15, in particular, serve to document some of this variation). Our analyses found that while the sites collect information on activities and short-run outcomes, the program lacks the detailed and nuanced metrics to measure longer-term effectiveness and determine necessary or optimal policy changes. In this chapter, we summarize our findings in terms of data reported by the sites for the ChalleNGe classes that occurred during 2015; models to explain ChalleNGe and assist the program in determining how to track progress; implications for future data collection; and data collection strategies.

## ChalleNGe Models and Current Metrics

This report is intended to serve two purposes. The first is to provide a snapshot of the program. The second is to begin the process of developing a richer set of metrics that describe the program's influence over a longer period and are tied to the program's overall mission of producing graduates with the values, skills, education, and self-discipline required to succeed as adults.

Our data collection describing the ChalleNGe classes of 2015 serves the first purpose (documenting program progress). Most of the metrics collected by ChalleNGe sites to date focus on inputs, activities, and outputs, with a few metrics of shorter-term outcomes. While these metrics form the basis of the program's longer-term impact, they do not serve to *measure* the longer-term impact. However, they do represent metrics of significant progress by cadets who took part in the program.

We place considerable focus on one existing metric, the TABE, which serves as a primary metric of academic progress among ChalleNGe participants. The test is generally quite appropriate for this purpose; indeed, USDOE requires that adult education programs use the TABE or similar approved assessments to track progress. However, the average grade equivalent scores reported by ChalleNGe sites do not indicate the number or proportion of cadets who have reached key benchmarks (and while such averages are widely used, they are also problematic from a measurement perspective; see Lindholm-Leary and Hargett, 2006). Fortunately, TABE-based benchmarks exist; we present two metrics that are linked to ChalleNGe-relevant outcomes: achieving a grade level of at least 9.0 (early high school) and achieving a grade level of at least 11.0 (late high school). We find that cadets who enter the program scoring at the middle school level or above are quite likely to achieve key benchmarks by graduation. If combined with a metric based on test score growth, reporting benchmarks achieved could provide a much more complete picture of ChalleNGe cadet performance, with little if any additional information collection required.

Our analysis of cost data provided by the sites indicates that while most ChalleNGe sites have somewhat similar costs in terms of average cost per graduate, a few sites have costs that are much higher. We explored several possible reasons for cost variation and found that, rather than differences in sites' ages or credentials awarded, size (number of graduates) is the driver of cost. Of course, we would expect costs to vary with the number of cadets, but sites that have fewer than 150 graduates per year have substantially higher costs than larger sites, while costs per graduate generally are quite similar at sites that have at least 150 graduates. This suggests that the fixed costs of running a ChalleNGe site dominate other costs in smaller sites. While these small sites are only responsible for about 6 percent of total costs, the data indicate that encouraging sites to attain a size of at least 150 graduates has the potential to improve cost-effectiveness.

To begin the process of improving program metrics and measuring longer-term impacts (the second purpose of this report), we have developed two tools: a TOC that describes the mechanisms underlying ChalleNGe and a program logic model that describes the relationships between resources, activities, and outcomes. These tools are useful for understanding the types of metrics and data collection efforts necessary to measure the longer-term impacts of the program (and for communicating program goals to stakeholders). In general, these models indicate that effectively linking aspects of the ChalleNGe program to longer-term outcomes will likely require additional data collection efforts. Adult education programs collect some relevant longer-term outcomes on their participants; some of their data collection strategies are relevant.

In many cases, adult education programs utilize existing administrative datasets (e.g., state UI datasets). While such datasets contain information relevant to ChalleNGe, the ChalleNGe program faces several barriers to such data collection strategies. These include mobility of participants, which many programs face, but a barrier that is especially relevant to ChalleNGe is related to the large number of sites in multiple states and the lack of formal linkages between ChalleNGe sites and relevant state departments. For these reasons, leveraging such administrative datasets represents a costly strategy in terms of establishing official data use agreements. Fortunately, the ChalleNGe sites have counselors in place to collect some data, which could be fine-tuned to represent better metrics. Finally, surveys of past cadets appear to represent a viable method of collecting additional information.

## Closing Thoughts, Next Steps

In closing, the ChalleNGe model appears well grounded in the existing literature on youth behavior and programs to positively influence youth behavior. However, collecting longer-term and/or more-detailed placement information is necessary to measure many of the long-term outcomes and impacts detailed in the program logic model (Chapter Three) and to determine the extent to which the ChalleNGe program is achieving its mission of producing graduates with the preparation, skills, and values necessary to succeed.

In future work, we will collect similar data to what is reported here, as well as additional data on the placements of cadets who attended a ChalleNGe program beginning in 2015. We will also collect additional data to document any time trends that could explain changes in the program's effectiveness. As part of this effort, we will work with individual programs to develop more comprehensive and cost-effective ways of collecting the required data. We will also continue to develop metrics linked to the ChalleNGe program's mission and specific strategies to assist the ChalleNGe sites in collecting data necessary to measure longer-term outcomes and impacts. We anticipate producing annual reports in 2017, 2018, and 2019. Between these reports we will also produce additional analyses on specific aspects of the ChalleNGe program, such as the number of young people lacking a high school diploma in states and metro areas (to increase the proportion potentially served by ChalleNGe). We will explore cost differences in more detail, and will analyze the impact of program-level factors such as classroom curricula, staffing ratios, and other factors on cadet success.

# Site-Specific Information and Data

In this appendix we provide additional information about the ChalleNGe sites and about the specific data collection effort that provided the information in this report.

## Site-Specific Information

Figure A.1 includes the start and stop dates for Classes 44 and 45, by site. This figure demonstrates the substantial variation in the timing of the classes across sites. Table A.1 includes entries for each site: program abbreviation (used in some of the figures), state, and program name. It demonstrates the distribution of sites across states.

**Figure A.1**
**2015–2016 Start and Stop Dates of ChalleNGe Sites, Classes 44 and 45**

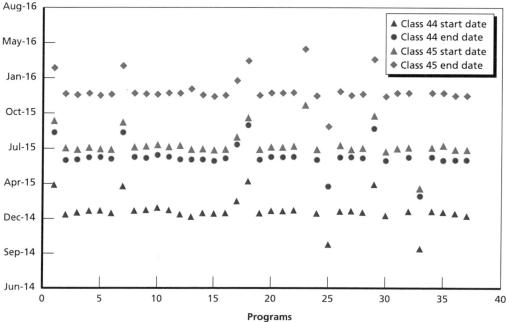

SOURCE: RAND analysis of data collected from ChalleNGe sites, Classes 44 and 45.
RAND *RR1848-A.1*

**Table A.1**
**National Guard Youth ChalleNGe: Program Abbreviation, State, Name**

| Program Abbreviation | State | Program Name |
|---|---|---|
| AK | Alaska | Alaska Military Youth Academy |
| AR | Arkansas | Arkansas Youth ChalleNGe |
| CA-LA | California | Sunburst Youth Academy |
| CA-SL | California | Grizzly Youth Academy |
| D.C. | District of Columbia | Capital Guardian Youth ChalleNGe Academy |
| FL | Florida | Florida Youth ChalleNGe Academy |
| GA-FG | Georgia | Fort Gordon Youth ChalleNGe Academy |
| GA-FS | Georgia | Fort Stewart Youth ChalleNGe Academy |
| HI-BP | Hawaii | Hawaii Youth ChalleNGe Academy at Barber's Point |
| HI-HI | Hawaii | Hawaii Youth ChalleNGe Academy at Hilo |
| ID | Idaho | Idaho Youth ChalleNGe Academy |
| IL | Illinois | Lincoln's ChalleNGe Academy |
| IN | Indiana | Hoosier Youth ChalleNGe Academy |
| KY-FK | Kentucky | Bluegrass ChalleNGe Academy |
| KY-HN | Kentucky | Appalachian ChalleNGe Program |
| LA-CB | Louisiana | Louisiana Youth ChalleNGe Program—Camp Beauregard |
| LA-CM | Louisiana | Louisiana Youth ChalleNGe Program—Camp Minden |
| LA-GL | Louisiana | Louisiana Youth ChalleNGe Program—Gillis Long |
| MD | Maryland | Freestate ChalleNGe Academy |
| MI | Michigan | Michigan Youth ChalleNGe Academy |
| MS | Mississippi | Mississippi Youth ChalleNGe Academy |
| MT | Montana | Montana Youth ChalleNGe Academy |
| NC-NL | North Carolina | Tarheel ChalleNGe Academy—New London |
| NC-S | North Carolina | Tarheel ChalleNGe Academy—Salemburg |
| NJ | New Jersey | New Jersey Youth ChalleNGe Academy |
| NM | New Mexico | New Mexico Youth ChalleNGe Academy |
| OK | Oklahoma | Thunderbird Youth Academy |
| OR | Oregon | Oregon Youth ChalleNGe Program |
| PR | Puerto Rico | Puerto Rico Youth ChalleNGe Academy |
| SC | South Carolina | South Carolina Youth ChalleNGe Academy |
| TX-E | Texas | Texas ChalleNGe Academy—East |
| TX-W | Texas | Texas ChalleNGe Academy—West |
| VA | Virginia | Virginia Commonwealth ChalleNGe Youth Academy |
| WA | Washington | Washington Youth Academy |
| WI | Wisconsin | Wisconsin ChalleNGe Academy |
| WV | West Virginia | Mountaineer ChalleNGe Academy |
| WY | Wyoming | Wyoming Cowboy ChalleNGe Academy |

# Tests of Adult Basic Education

The TABE is one of the three most commonly used assessments in ABE/ASE programs that serve out-of-school youth (at least 16 years of age) and adults who have not yet demonstrated the skills and/or competencies required for obtaining a high school diploma or a high school equivalency degree. Such programs are administered by a variety of institutions, such as public school districts, community colleges, nonprofit agencies, or volunteer organizations ("Adult Education—CalEdFacts," 2017). The Office of Career, Technical, and Adult Education, a subdivision of the USDOE, oversees the federal funding and regulations that guide state ABE/ASE programs.

Among the requirements governing programs that use federal education funds is a requirement to annually report enrollment numbers, demographics, annual learning gains, and job placement statistics of participating individuals (34 C.F.R. §§ 461–463, 1998). The USDOE has approved the use of the TABE, CASAS, Massachusetts Adult Proficiency Test (MAPT), or General Assessment of Instructional Needs (GAIN) for the reporting of learning gains in ABE/ASE programs of native English speakers (80 F.R. § 48304, 2015). Despite this, there is little information available linking the TABE to other assessments or other outcomes of interest.

The TABE is offered in both paper and online formats using multiple-choice questions. The complete battery of core test areas (195 questions) is suggested to take three hours, while the survey exam format of 100 test questions can be completed in approximately an hour and a half. The four core test areas on both the survey and complete battery exams include: Reading, Mathematics Computation, Applied Mathematics, and Language (Language Mechanics, Vocabulary, and Spelling are optional areas that can be assessed).

The TABE serves two assessment purposes: as a formative/diagnostic assessment and as a summative assessment ("TABE Tests of Adult Basic Education," n.d.). Formative/diagnostic assessments measure an individual's current education level and identify topics that an individual needs additional support in or instruction on, which helps guide instructors on how to focus instructional efforts for individual learners. A summative assessment identifies the performance level or how much an individual has learned (i.e., learning gains) following instruction or educational services.

To allow for the assessment of learning gains, the TABE offers two forms for each of the five levels of the exam. These five levels correspond to grade-span ranges (K–1, 2–3, 4–5, 6–8, and 9–12). The TABE provides a brief "Locator" assessment to identify the appropriate level of exam an individual should be given. The two forms of each exam level allow a student to take the first form of, for example, the exam for grades 6–8 upon entry into a program and then the second form when measuring learning gains. The same student could also be given the first form of the exam for grades 9–12 at the point of measuring learning gains if the Locator assessment identifies this exam level as appropriate for that individual.

Federal guidelines on how much time must pass between test administrations (from pretest/formative assessment to the posttest/summative assessment) do not exist (USDOE OCTAE Department of Adult Education Literacy, 2016). The Massachusetts Department of Elementary and Secondary Education (2015) suggests at least 65 instructional hours be delivered prior to administering the second form of TABE as a summative assessment. New York recognizes that program hours differ across institutions and that test intervals should be set accordingly, but a minimum of 40 program hours between pre- and posttest administrations is recommended (New York State Education Department, 2015). According to the City Colleges of Chicago (2015), the TABE test publisher's testing guidelines require a minimum of 40 instructional hours between pre- and posttesting.

The TABE offers three types of scoring information: a number of correct responses, a scale score, and a grade equivalent score.[1] Limited research is available to map TABE scores onto similar assessments or assessments of related interest, and much of the documentation available comes from community colleges or state departments of education that use test scores from the aforementioned exams to appropriately place students into academic courses. For example, the NRS, the accountability system for ABE/ASE programs, provides test score ranges that link TABE scale scores and grade equivalents to scale score ranges on the CASAS, GAIN, and MAPT (National Reporting Service for Adult Education, 2015). Tables B.1 and B.2 provide crosswalks between TABE scores and other outcomes relevant to the ChalleNGe program. The tables come from separate documents: the first document includes actual pass rates, and the second document includes estimated probabilities of passing the GED. In general, the information across the two sources is quite consistent.

---

[1]   A grade equivalent score (e.g., 5.9) is commonly misinterpreted as reflecting mastery of the standards in a particular grade. Receiving a grade equivalent score of 5.9 suggests that in current test administration, the student's performance was similar to that of an individual performing at the 50th percentile of students who were in the ninth month of fifth grade. A 6.9 suggests the individual is performing similar to students in the 50th percentile at the end of sixth grade (the ninth month of sixth grade).

**Table B.1**
**TABE Grade Level Equivalents, CASAS Scores, and GED Passing Rates**

| TABE Level | Grade Level | TABE Scale Scores Reading | TABE Scale Scores Math | TABE Scale Scores Language | Related CASAS Scale Score | GED Read Pass Rate | GED | GED Math Pass Rate |
|---|---|---|---|---|---|---|---|---|
| Beginning adult basic | 0.0–1.9 | <=367 | <=313 | <=392 | 181–200 | — | — | — |
| Beginning basic | 2.0–3.9 | 368–460 | 314–441 | 393–490 | 201–210 | — | — | — |
| Low intermediate basic | 4.0–5.9 | 461–517 | 442–505 | 491–523 | 211–220 | — | — | — |
| High intermediate basic | 6.0–8.9 | 518–566 | 506–565 | 524–559 | 221–235 | — | — | — |
| Low adult secondary | 9.0–10.9 | 567–595 | 566–594 | 560–585 | 236–245 | 75% | 70% | 90% |
| High adult secondary | 11.0–12.9 | >=596 | >=595 | >=586 | >=246 | 89% | 85% | 97% |

SOURCE: CASAS (2003).
NOTE: GED pass rates were associated with CASAS scores.

**Table B.2**
**TABE Grade Level Equivalents, CASAS Scores, and Predicted GED Readiness Levels**

| TABE Level | TABE Grade Level Equivalents | TABE Scale Scores Reading | TABE Scale Score Total Math | TABE, Scale Score Lang | Related CASAS Scale Score | GED "Not Likely to Pass" | GED "Too Close to Call" | GED "Likely to Pass" |
|---|---|---|---|---|---|---|---|---|
| Beginning adult basic | 0.0–1.9 | <=367 | <=313 | <=392 | 181–200 | 100% | 0% | 0% |
| Beginning basic | 2.0–3.9 | 368–460 | 314–441 | 393–490 | 201–210 | 100% | 0% | 0% |
| Low intermediate basic | 4.0–5.9 | 461–517 | 442–505 | 491–523 | 211–220 | 60% | 36% | 4% |
| High intermediate basic | 6.0–8.9 | 518–566 | 506–565 | 524–559 | 221–235 | 22% | 55% | 23% |
| Low adult secondary | 9.0–10.9 | 567–595 | 566–594 | 560–585 | 236–245 | 8% | 34% | 58% |
| High adult secondary | 11.0–12.9 | >=596 | >=595 | >=586 | >=246 | 3% | 14% | 83% |

SOURCE: CASAS (2016).
NOTE: GED Ready Scores for students below 211 are all grouped into one category and identified as not ready for the GED.

# Detailed Information by ChalleNGe Site

In this appendix we present detailed information for each site. This information serves to document program progress and support the annual report to Congress. The tables allow the reader to see all of a given program's information at once, thereby offering a more detailed understanding of each program.

The sites are listed alphabetically by state or territory name. Each table includes metrics of the number and type of staff, total funding (in 2015), as well as the numbers of cadets who applied, graduated, and received various credentials. The tables also include data related to several of the core components—service to community (and calculated values based on local labor market conditions), gains on specific physical fitness tests, as well as the numbers of cadets registered to vote or registered for Selective Service. Finally, the tables include information about postgraduation placement (although there is no information on Class 45 12-month placement rates because fewer than 12 months have passed since graduation for this group).

In a few cases the data reported by the sites did not meet quality assurance standards; in these cases we have elected not to report the data. An example of such a case would be a site that reported highly unlikely run times. In other cases, the data had small anomalies; when small anomalies were present, we elected to report the data. An example of such an anomaly would be a site that indicated they contacted one or two more cadets than the number who graduated.

**Table C.1**
**Alaska Profile**

| ALASKA MILITARY YOUTH ACADEMY, ESTABLISHED 1994 |
|---|

Graduates since inception: 4,800 | Program type: Credit Recovery, High School Diploma, GED or HiSET

### Staffing

|  | Instructional | Cadre | Administrative | Other | Total |
|---|---|---|---|---|---|
| Full-time | 7 | 28 | 9 | 18 | 62 |
| Part-time | 0 | 0 | 0 | 1 | 1 |

### Funding

|  | Federal Funding | State Funding |
|---|---|---|
| Classes 44 and 45 | $3,715,000 | $1,238,000 |

### Residential Performance

|  | Dates | Target | Applied | Graduated | Received GED/ HiSET | Received HS Credits | Received HS Diploma |
|---|---|---|---|---|---|---|---|
| Class 44 | Apr. 2015–Aug. 2015 | 144 | 232 | 154 | 53 | 150 | 25 |
| Class 45 | Oct. 2015–Feb. 2016 | 144 | 184 | 118 | 67 | 117 | 14 |

### Physical Fitness

|  | Curl-Ups | | Push-Ups | | 1-Mile Run | |
|---|---|---|---|---|---|---|
|  | Initial | Final | Initial | Final | Initial | Final |
| Class 44 | 37.9 | 46.6 | * | * | 10:27 | 07:47 |
| Class 45 | 32.5 | 47.3 | * | * | 10:53 | 07:51 |

### Responsible Citizenship

|  | Voting | | Selective Service | |
|---|---|---|---|---|
|  | Eligible | Registered | Eligible | Registered |
| Class 44 | 44 | 44 | 27 | 27 |
| Class 45 | 32 | 32 | 26 | 26 |

### Service to Community

|  | Hours of Service | Dollar Value/Hr | Total Value |
|---|---|---|---|
| Class 44 | 10,984 | $27.51 | $302,170 |
| Class 45 | 6,809 | $27.51 | $187,316 |

### Postresidential Performance Status

|  | Graduated | Contacted | Placed | Education | Employment | Military | Other |
|---|---|---|---|---|---|---|---|
| Class 44 |  |  |  |  |  |  |  |
| Month 1 | 154 | 154 | 139 | 99 | 23 | 1 | 16 |
| Month 6 | 154 | 154 | 133 | 102 | 27 | 1 | 3 |
| Month 12 | 154 | 154 | 112 | 46 | 50 | 6 | 10 |
| Class 45 |  |  |  |  |  |  |  |
| Month 1 | 118 | 118 | 110 | 86 | 15 | 0 | 9 |
| Month 6 | 118 | 118 | 105 | 34 | 64 | 1 | 6 |

* Did not report

**Table C.2**
**Arkansas Profile**

| ARKANSAS YOUTH CHALLENGE, ESTABLISHED 1993 |
|---|

Graduates since inception: 3,396                                     Program type: GED or HiSET

### Staffing

|  | Instructional | Cadre | Administrative | Other | Total |
|---|---|---|---|---|---|
| Full-time | 4 | 18 | 15 | 0 | 37 |
| Part-time | 0 | 0 | 0 | 0 | 0 |

### Funding

|  | Federal Funding | State Funding |
|---|---|---|
| Classes 44 and 45 | $2,300,000 | $767,000 |

### Residential Performance

|  | Dates | Target | Applied | Graduated | Received GED/ HiSET | Received HS Credits | Received HS Diploma |
|---|---|---|---|---|---|---|---|
| Class 44 | Jan. 2015–June 2015 | 100 | 195 | 96 | 24 | ~ | ~ |
| Class 45 | July 2015–Dec. 2015 | 100 | 233 | 97 | 15 | ~ | ~ |

### Physical Fitness

|  | Curl-Ups | | Push-Ups | | 1-Mile Run | |
|---|---|---|---|---|---|---|
|  | Initial | Final | Initial | Final | Initial | Final |
| Class 44 | 29.3 | 29.6 | 22.8 | 44.3 | 11:01 | 11:20 |
| Class 45 | 29.3 | 33.8 | 31.2 | 44.9 | 10:42 | 09:58 |

### Responsible Citizenship

|  | Voting | | Selective Service | |
|---|---|---|---|---|
|  | Eligible | Registered | Eligible | Registered |
| Class 44 | 29 | 29 | 59 | 59 |
| Class 45 | 28 | 28 | 55 | 55 |

### Service to Community

|  | Hours of Service | Dollar Value/Hr | Total Value |
|---|---|---|---|
| Class 44 | 7,213 | $19.14 | $138,057 |
| Class 45 | 7,071 | $19.14 | $135,339 |

### Postresidential Performance Status

|  | Graduated | Contacted | Placed | Education | Employment | Military | Other |
|---|---|---|---|---|---|---|---|
| Class 44 |  |  |  |  |  |  |  |
| Month 1 | 96 | 98 | 35 | 19 | 28 | 2 | 9 |
| Month 6 | 96 | 98 | 55 | 52 | 31 | 3 | 9 |
| Month 12 | 96 | 98 | 60 | 41 | 35 | 3 | 5 |
| Class 45 |  |  |  |  |  |  |  |
| Month 1 | 97 | 97 | 64 | 58 | 14 | 0 | 2 |
| Month 6 | 97 | 97 | 69 | 54 | 40 | 7 | 4 |

~ Does not award

**Table C.3**
**California, Sunburst Youth Academy Profile**

### SUNBURST YOUTH ACADEMY, ESTABLISHED 2008

| Graduates since inception: 2,441 | Program type: High School Diploma, GED, Credit Recovery |
|---|---|

#### Staffing

|  | Instructional | Cadre | Administrative | Other | Total |
|---|---|---|---|---|---|
| Full-time | 22 | 26 | 3 | 20 | 71 |
| Part-time | 0 | 0 | 0 | 0 | 0 |

#### Funding

|  | Federal Funding | State Funding |
|---|---|---|
| Classes 44 and 45 | $5,400,000 | $1,950,000 |

#### Residential Performance

|  | Dates | Target | Applied | Graduated | Received GED/ HiSET | Received HS Credits | Received HS Diploma |
|---|---|---|---|---|---|---|---|
| Class 44 | Jan. 2015–June 2015 | 180 | 255 | 193 | * | 193 | 19 |
| Class 45 | July 2015–Dec. 2015 | 180 | 208 | 184 | * | 184 | 18 |

#### Physical Fitness

|  | Curl-Ups | | Push-Ups | | 1-Mile Run | |
|---|---|---|---|---|---|---|
|  | Initial | Final | Initial | Final | Initial | Final |
| Class 44 | 30.4 | 44.8 | 25.1 | 56.8 | 08:19 | 07:02 |
| Class 45 | 25.0 | 40.9 | 20.2 | 46.8 | 09:34 | 07:55 |

#### Responsible Citizenship

|  | Voting | | Selective Service | |
|---|---|---|---|---|
|  | Eligible | Registered | Eligible | Registered |
| Class 44 | 42 | 42 | 43 | 43 |
| Class 45 | 34 | 34 | 34 | 34 |

#### Service to Community

|  | Hours of Service | Dollar Value/Hr | Total Value |
|---|---|---|---|
| Class 44 | 8,695 | $27.59 | $239,895 |
| Class 45 | 8,423 | $27.59 | $232,377 |

#### Postresidential Performance Status

|  | Graduated | Contacted | Placed | Education | Employment | Military | Other |
|---|---|---|---|---|---|---|---|
| Class 44 |  |  |  |  |  |  |  |
| Month 1 | 193 | 180 | 167 | 156 | 36 | 1 | 5 |
| Month 6 | 193 | 174 | 163 | 145 | 63 | 4 | 2 |
| Month 12 | 193 | 173 | 170 | 121 | 78 | 8 | 9 |
| Class 45 |  |  |  |  |  |  |  |
| Month 1 | 184 | 165 | 151 | 147 | 17 | 0 | 3 |
| Month 6 | 184 | 164 | 156 | 141 | 54 | 5 | 8 |

* Did not report

**Table C.4**
**California, Grizzly Youth Academy Profile**

| GRIZZLY YOUTH ACADEMY, ESTABLISHED 1998 |
|---|

Graduates since inception: 5,096

Program type: Credit Recovery, High School Diploma, GED or HiSET

### Staffing

|  | Instructional | Cadre | Administrative | Other | Total |
|---|---|---|---|---|---|
| Full-time | 11 | 32 | 23 | 0 | 66 |
| Part-time | 0 | 0 | 0 | 0 | 0 |

### Funding

|  | Federal Funding | State Funding |
|---|---|---|
| Classes 44 and 45 | $5,500,000 | $4,188,857 |

### Residential Performance

|  | Dates | Target | Applied | Graduated | Received GED/ HiSET | Received HS Credits | Received HS Diploma |
|---|---|---|---|---|---|---|---|
| Class 44 | Jan. 2015–June 2015 | 180 | 247 | 191 | 5 | 189 | 54 |
| Class 45 | July 2015–Dec. 2015 | 185 | 304 | 208 | 4 | 208 | 60 |

### Physical Fitness

|  | Curl-Ups | | Push-Ups | | 1-Mile Run | |
|---|---|---|---|---|---|---|
|  | Initial | Final | Initial | Final | Initial | Final |
| Class 44 | 30.6 | 43.7 | 20.3 | 36.2 | 09:21 | 07:33 |
| Class 45 | 30.7 | 42.9 | 20.8 | 35.4 | 09:56 | 07:48 |

### Responsible Citizenship

|  | Voting | | Selective Service | |
|---|---|---|---|---|
|  | Eligible | Registered | Eligible | Registered |
| Class 44 | 45 | 45 | 45 | 45 |
| Class 45 | 43 | 43 | 43 | 43 |

### Service to Community

|  | Hours of Service | Dollar Value/Hr | Total Value |
|---|---|---|---|
| Class 44 | 20,024 | $27.59 | $552,462 |
| Class 45 | 13,056 | $27.59 | $360,215 |

### Postresidential Performance Status

|  | Graduated | Contacted | Placed | Education | Employment | Military | Other |
|---|---|---|---|---|---|---|---|
| Class 44 |  |  |  |  |  |  |  |
| Month 1 | 191 | 191 | 140 | 119 | 58 | 1 | 14 |
| Month 6 | 191 | 191 | 174 | 154 | 126 | 8 | 14 |
| Month 12 | 191 | 191 | 158 | 131 | 119 | 15 | 11 |
| Class 45 |  |  |  |  |  |  |  |
| Month 1 | 208 | 208 | 193 | 167 | 57 | 1 | 10 |
| Month 6 | 208 | 208 | 191 | 160 | 107 | 7 | 12 |

**Table C.5**
**District of Columbia Profile**

| CAPITAL GUARDIAN YOUTH CHALLENGE ACADEMY, ESTABLISHED 2007 |
| --- |

| Graduates since inception: 460 | Program type: GED or HiSET |
| --- | --- |

**Staffing**

|  | Instructional | Cadre | Administrative | Other | Total |
| --- | --- | --- | --- | --- | --- |
| Full-time | 6 | 22 | 28 | 0 | 56 |
| Part-time | 0 | 0 | 0 | 0 | 0 |

**Funding**

|  | Federal Funding | State Funding |
| --- | --- | --- |
| Classes 44 and 45 | $2,700,000 | $1,880,930 |

**Residential Performance**

|  | Dates | Target | Applied | Graduated | Received GED/ HiSET | Received HS Credits | Received HS Diploma |
| --- | --- | --- | --- | --- | --- | --- | --- |
| Class 44 | Jan. 2015–June 2015 | 100 | 65 | 36 | * | ~ | 9 |
| Class 45 | July 2015–Dec. 2015 | 100 | 81 | 57 | * | ~ | 21 |

**Physical Fitness**

|  | Curl-Ups | | Push-Ups | | 1-Mile Run | |
| --- | --- | --- | --- | --- | --- | --- |
|  | Initial | Final | Initial | Final | Initial | Final |
| Class 44 | 26.5 | 34.2 | 16.1 | 28.2 | 10:50 | 09:15 |
| Class 45 | 26.1 | 30.9 | 19.4 | 25.1 | 13:34 | 11:44 |

**Responsible Citizenship**

|  | Voting | | Selective Service | |
| --- | --- | --- | --- | --- |
|  | Eligible | Registered | Eligible | Registered |
| Class 44 | 11 | 11 | 7 | 7 |
| Class 45 | 8 | 8 | 7 | 7 |

**Service to Community**

|  | Hours of Service | Dollar Value/Hr | Total Value |
| --- | --- | --- | --- |
| Class 44 | 2,150 | $38.77 | $83,356 |
| Class 45 | 1,505 | $38.77 | $58,349 |

**Postresidential Performance Status**

|  | Graduated | Contacted | Placed | Education | Employment | Military | Other |
| --- | --- | --- | --- | --- | --- | --- | --- |
| Class 44 |  |  |  |  |  |  |  |
| Month 1 | 36 | 50 | 20 | 10 | 10 | 0 | 0 |
| Month 6 | 36 | 50 | 7 | 3 | 3 | 1 | 0 |
| Month 12 | 36 | 50 | 16 | 7 | 8 | 1 | 0 |
| Class 45 |  |  |  |  |  |  |  |
| Month 1 | 57 | 35 | 17 | 13 | 5 | 0 | 5 |
| Month 6 | 57 | 32 | 26 | 13 | 10 | 0 | 3 |

* Did not report

~ Does not award

**Table C.6**
**Florida Profile**

| FLORIDA YOUTH CHALLENGE ACADEMY, ESTABLISHED 2001 |
|---|

| Graduates since inception: 4,030 | Program type: GED or HiSET, Credit Recovery |
|---|---|

### Staffing

| | Instructional | Cadre | Administrative | Other | Total |
|---|---|---|---|---|---|
| Full-time | 7 | 41 | 9 | 27 | 84 |
| Part-time | 0 | 0 | 0 | 0 | 0 |

### Funding

| | Federal Funding | State Funding |
|---|---|---|
| Classes 44 and 45 | $3,200,000 | $1,300,000 |

### Residential Performance

| | Dates | Target | Applied | Graduated | Received GED/ HiSET | Received HS Credits | Received HS Diploma |
|---|---|---|---|---|---|---|---|
| Class 44 | Jan. 2015–June 2015 | 150 | 245 | 165 | 65 | 165 | ~ |
| Class 45 | July 2015–Dec. 2015 | 150 | 265 | 167 | 73 | 34 | ~ |

### Physical Fitness

| | Curl-Ups | | Push-Ups | | 1-Mile Run | |
|---|---|---|---|---|---|---|
| | Initial | Final | Initial | Final | Initial | Final |
| Class 44 | 35.9 | 71.8 | 19.9 | 36.0 | 10:01 | 07:55 |
| Class 45 | 43.9 | 67.0 | 16.2 | 37.8 | 10:18 | 08:06 |

### Responsible Citizenship

| | Voting | | Selective Service | |
|---|---|---|---|---|
| | Eligible | Registered | Eligible | Registered |
| Class 44 | 43 | 43 | 49 | 49 |
| Class 45 | 54 | 54 | 43 | 43 |

### Service to Community

| | Hours of Service | Dollar Value/Hr | Total Value |
|---|---|---|---|
| Class 44 | 8,248 | $22.08 | $182,116 |
| Class 45 | 8,038 | $22.08 | $177,479 |

### Postresidential Performance Status

| | Graduated | Contacted | Placed | Education | Employment | Military | Other |
|---|---|---|---|---|---|---|---|
| Class 44 | | | | | | | |
| Month 1 | 165 | 165 | 130 | 32 | 90 | 0 | 8 |
| Month 6 | 165 | 165 | 124 | 40 | 78 | 1 | 5 |
| Month 12 | 165 | 39 | 29 | 3 | 23 | 0 | 3 |
| Class 45 | | | | | | | |
| Month 1 | 167 | 167 | 124 | 49 | 66 | 1 | 8 |
| Month 6 | 167 | 84 | 38 | 4 | 32 | 2 | 0 |

~ Does not award

**Table C.7**
**Georgia, Fort Gordon Youth Academy Profile**

| FORT GORDON YOUTH CHALLENGE ACADEMY, ESTABLISHED 2000 |
|---|

Graduates since inception: 5,410

Program type: GED, High School Diploma, Credit Recovery

**Staffing**

| | Instructional | Cadre | Administrative | Other | Total |
|---|---|---|---|---|---|
| Full-time | 9 | 44 | 8 | 31 | 92 |
| Part-time | 0 | 5 | 3 | 10 | 18 |

**Funding**

| | Federal Funding | State Funding |
|---|---|---|
| Classes 44 and 45 | $5,140,009 | $1,713,337 |

**Residential Performance**

| | Dates | Target | Applied | Graduated | Received GED/HiSET | Received HS Credits | Received HS Diploma |
|---|---|---|---|---|---|---|---|
| Class 44 | Mar. 2015–Aug. 2015 | 212 | 533 | 192 | 79 | * | * |
| Class 45 | Sept. 2015–Mar. 2016 | 213 | 389 | 182 | 113 | 7 | 5 |

**Physical Fitness**

| | Curl-Ups | | Push-Ups | | 1-Mile Run | |
|---|---|---|---|---|---|---|
| | Initial | Final | Initial | Final | Initial | Final |
| Class 44 | * | * | * | * | * | * |
| Class 45 | 40.3 | 42.9 | 32.5 | 41.1 | 09:11 | 08:42 |

**Responsible Citizenship**

| | Voting | | Selective Service | |
|---|---|---|---|---|
| | Eligible | Registered | Eligible | Registered |
| Class 44 | 56 | 56 | 92 | 92 |
| Class 45 | 62 | 62 | 83 | 83 |

**Service to Community**

| | Hours of Service | Dollar Value/Hr | Total Value |
|---|---|---|---|
| Class 44 | 11,342 | $23.80 | $269,940 |
| Class 45 | 10,159 | $23.80 | $241,784 |

**Postresidential Performance Status**

| | Graduated | Contacted | Placed | Education | Employment | Military | Other |
|---|---|---|---|---|---|---|---|
| Class 44 | | | | | | | |
| Month 1 | 192 | 188 | 132 | 44 | 62 | 0 | 19 |
| Month 6 | 192 | 188 | 130 | 49 | 71 | 5 | 12 |
| Month 12 | 192 | 188 | 113 | 30 | 76 | 8 | 8 |
| Class 45 | | | | | | | |
| Month 1 | 182 | 181 | 124 | 41 | 61 | 0 | 25 |
| Month 6 | 182 | 181 | 128 | 23 | 87 | 3 | 12 |

* Did not report

**Table C.8**
**Georgia, Fort Stewart Youth Academy Profile**

| FORT STEWART YOUTH CHALLENGE ACADEMY, ESTABLISHED 1993 |
|---|

| Graduates since inception: 8,878 | Program type: GED, High School Diploma, High School Credit Recovery |
|---|---|

**Staffing**

| | Instructional | Cadre | Administrative | Other | Total |
|---|---|---|---|---|---|
| Full-time | 7 | 44 | 47 | 0 | 98 |
| Part-time | 1 | 11 | 4 | 0 | 16 |

**Funding**

| | Federal Funding | State Funding |
|---|---|---|
| Classes 44 and 45 | $5,062,500 | $1,717,617 |

**Residential Performance**

| | Dates | Target | Applied | Graduated | Received GED/ HiSET | Received HS Credits | Received HS Diploma |
|---|---|---|---|---|---|---|---|
| Class 44 | Jan. 2015–June 2015 | 213 | 374 | 172 | 112 | 18 | 18 |
| Class 45 | July 2015–Dec. 2015 | 212 | 395 | 214 | 148 | 33 | 32 |

**Physical Fitness**

| | Curl-Ups | | Push-Ups | | 1-Mile Run | |
|---|---|---|---|---|---|---|
| | Initial | Final | Initial | Final | Initial | Final |
| Class 44 | 40.2 | 50.5 | * | * | 09:16 | 08:36 |
| Class 45 | 8.9 | 41.1 | * | * | 08:50 | 09:05 |

**Responsible Citizenship**

| | Voting | | Selective Service | |
|---|---|---|---|---|
| | Eligible | Registered | Eligible | Registered |
| Class 44 | 54 | 54 | 49 | 49 |
| Class 45 | 67 | 67 | 57 | 57 |

**Service to Community**

| | Hours of Service | Dollar Value/Hr | Total Value |
|---|---|---|---|
| Class 44 | 10,258 | $23.80 | $244,140 |
| Class 45 | 12,130 | $23.80 | $288,682 |

**Postresidential Performance Status**

| | Graduated | Contacted | Placed | Education | Employment | Military | Other |
|---|---|---|---|---|---|---|---|
| Class 44 | | | | | | | |
| Month 1 | 172 | 171 | 155 | 75 | 72 | 2 | 21 |
| Month 6 | 172 | 171 | 156 | 78 | 74 | 2 | 15 |
| Month 12 | 172 | 171 | 164 | 65 | 86 | 4 | 16 |
| Class 45 | | | | | | | |
| Month 1 | 214 | 214 | 175 | 110 | 69 | 2 | 1 |
| Month 6 | 214 | 214 | 116 | * | * | * | * |

* Did not report

**Table C.9**
**Hawaii, ChalleNGe Academy at Barber's Point Profile**

| HAWAII YOUTH CHALLENGE ACADEMY AT BARBER'S POINT, ESTABLISHED 1993 |
|---|

Graduates since inception: 3,847                    Program type: Other

### Staffing

|  | Instructional | Cadre | Administrative | Other | Total |
|---|---|---|---|---|---|
| Full-time | 6 | 26 | 20 | 2 | 54 |
| Part-time | 0 | 0 | 0 | 0 | 0 |

### Funding

|  | Federal Funding | State Funding |
|---|---|---|
| Classes 44 and 45 | $2,400,000 | $800,000 |

### Residential Performance

|  | Dates | Target | Applied | Graduated | Received GED/ HiSET | Received HS Credits | Received HS Diploma |
|---|---|---|---|---|---|---|---|
| Class 44 | Jan. 2015–June 2015 | 100 | 195 | 81 | 81 | ~ | ~ |
| Class 45 | July 2015–Dec. 2015 | 100 | 224 | 129 | 129 | ~ | ~ |

### Physical Fitness

|  | Curl-Ups | | Push-Ups | | 1-Mile Run | |
|---|---|---|---|---|---|---|
|  | Initial | Final | Initial | Final | Initial | Final |
| Class 44 | 34.9 | 47.5 | 39.5 | 53.9 | 11:23 | 09:33 |
| Class 45 | 31.2 | 51.7 | 34.0 | 58.7 | 11:06 | 08:13 |

### Responsible Citizenship

|  | Voting | | Selective Service | |
|---|---|---|---|---|
|  | Eligible | Registered | Eligible | Registered |
| Class 44 | 44 | 44 | 39 | 39 |
| Class 45 | 32 | 32 | 20 | 20 |

### Service to Community

|  | Hours of Service | Dollar Value/Hr | Total Value |
|---|---|---|---|
| Class 44 | 9,212 | $23.33 | $214,904 |
| Class 45 | 15,401 | $23.33 | $359,294 |

### Postresidential Performance Status

|  | Graduated | Contacted | Placed | Education | Employment | Military | Other |
|---|---|---|---|---|---|---|---|
| Class 44 |  |  |  |  |  |  |  |
| Month 1 | 81 | 82 | 53 | 0 | 29 | 0 | 24 |
| Month 6 | 81 | 82 | 76 | 4 | 37 | 2 | 36 |
| Month 12 | 81 | 82 | 75 | 2 | 29 | 4 | 43 |
| Class 45 |  |  |  |  |  |  |  |
| Month 1 | 129 | 129 | 90 | 10 | 44 | 0 | 38 |
| Month 6 | 129 | 129 | 114 | 5 | 29 | 2 | 79 |

~ Does not award

**Table C.10**
**Hawaii, Youth Academy at Hilo Profile**

### HAWAII YOUTH CHALLENGE ACADEMY AT HILO, ESTABLISHED 2011

| Graduates since inception: 518 | | | | Program type: High School Diploma | | | |
|---|---|---|---|---|---|---|---|

**Staffing**

| | Instructional | Cadre | Administrative | Other | Total |
|---|---|---|---|---|---|
| Full-time | 5 | 26 | 21 | 0 | 52 |
| Part-time | 0 | 0 | 0 | 0 | 0 |

**Funding**

| | Federal Funding | State Funding |
|---|---|---|
| Classes 44 and 45 | $2,100,000 | $700,000 |

**Residential Performance**

| | Dates | Target | Applied | Graduated | Received GED/ HiSET | Received HS Credits | Received HS Diploma |
|---|---|---|---|---|---|---|---|
| Class 44 | Jan. 2015–June 2015 | 100 | 108 | 60 | ~ | ~ | * |
| Class 45 | July 2015–Dec. 2015 | 100 | 112 | 70 | ~ | ~ | 70 |

**Physical Fitness**

| | Curl-Ups | | Push-Ups | | 1-Mile Run | |
|---|---|---|---|---|---|---|
| | Initial | Final | Initial | Final | Initial | Final |
| Class 44 | 41.8 | 63.4 | 50.5 | 70.7 | 09:39 | 07:42 |
| Class 45 | 52.5 | 73.3 | 51.8 | 71.5 | 09:30 | 08:09 |

**Responsible Citizenship**

| | Voting | | Selective Service | |
|---|---|---|---|---|
| | Eligible | Registered | Eligible | Registered |
| Class 44 | 32 | 32 | 45 | 45 |
| Class 45 | 11 | 11 | 38 | 38 |

**Service to Community**

| | Hours of Service | Dollar Value/Hr | Total Value |
|---|---|---|---|
| Class 44 | 6,708 | $23.33 | $156,498 |
| Class 45 | 9,402 | $23.33 | $219,349 |

**Postresidential Performance Status**

| | Graduated | Contacted | Placed | Education | Employment | Military | Other |
|---|---|---|---|---|---|---|---|
| Class 44 | | | | | | | |
| Month 1 | 60 | 21 | 2 | 0 | 1 | 0 | 1 |
| Month 6 | 60 | 15 | 6 | 1 | 5 | 0 | 0 |
| Month 12 | 60 | 5 | 0 | 0 | 0 | 0 | 0 |
| Class 45 | | | | | | | |
| Month 1 | 70 | 25 | 1 | 0 | 0 | 0 | 1 |
| Month 6 | 70 | 9 | 0 | 0 | 0 | 0 | 0 |

* Did not report

~ Does not award

**Table C.11**
**Idaho Profile**

### IDAHO YOUTH CHALLENGE ACADEMY, ESTABLISHED 2014

Graduates since inception: 333

Program type: High School Diploma,
Credit Recovery, GED or HiSET

#### Staffing

|  | Instructional | Cadre | Administrative | Other | Total |
|---|---|---|---|---|---|
| Full-time | 6 | 21 | 6 | 12 | 45 |
| Part-time | 0 | 0 | 0 | 7 | 7 |

#### Funding

|  | Federal Funding | State Funding |
|---|---|---|
| Classes 44 and 45 | $2,400,000 | $800,000 |

#### Residential Performance

|  | Dates | Target | Applied | Graduated | Received GED/ HiSET | Received HS Credits | Received HS Diploma |
|---|---|---|---|---|---|---|---|
| Class 44 | Jan. 2015–June 2015 | 100 | 108 | 81 | 7 | 81 | 2 |
| Class 45 | July 2015–Dec. 2015 | 100 | 127 | 101 | 12 | 101 | 12 |

#### Physical Fitness

|  | Curl-Ups | | Push-Ups | | 1-Mile Run | |
|---|---|---|---|---|---|---|
|  | Initial | Final | Initial | Final | Initial | Final |
| Class 44 | 54.4 | 68.7 | 33.0 | 45.1 | 10:33 | 08:14 |
| Class 45 | 49.8 | 70.1 | 25.1 | 44.7 | 10:12 | 07:51 |

#### Responsible Citizenship

|  | Voting | | Selective Service | |
|---|---|---|---|---|
|  | Eligible | Registered | Eligible | Registered |
| Class 44 | 13 | 13 | 18 | 18 |
| Class 45 | 19 | 19 | 31 | 31 |

#### Service to Community

|  | Hours of Service | Dollar Value/Hr | Total Value |
|---|---|---|---|
| Class 44 | 4,904 | $20.97 | $102,837 |
| Class 45 | 4,474 | $20.97 | $93,809 |

#### Postresidential Performance Status

|  | Graduated | Contacted | Placed | Education | Employment | Military | Other |
|---|---|---|---|---|---|---|---|
| Class 44 |  |  |  |  |  |  |  |
| Month 1 | 81 | 76 | 52 | 28 | 23 | 0 | 1 |
| Month 6 | 81 | 74 | 66 | 54 | 12 | 0 | 0 |
| Month 12 | 81 | 72 | 64 | 42 | 19 | 1 | 2 |
| Class 45 |  |  |  |  |  |  |  |
| Month 1 | 101 | 89 | 76 | 66 | 8 | 0 | 2 |
| Month 6 | 101 | 90 | 84 | 58 | 19 | 3 | 4 |

**Table C.12**
**Illinois Profile**

### LINCOLN'S CHALLENGE ACADEMY, ESTABLISHED 1993

Graduates since inception: 14,598                                    Program type: GED or HiSET

#### Staffing

|            | Instructional | Cadre | Administrative | Other | Total |
|------------|---------------|-------|----------------|-------|-------|
| Full-time  | 7             | 41    | 35             | 19    | 102   |
| Part-time  | 0             | 0     | 0              | 0     | 0     |

#### Funding

|                    | Federal Funding | State Funding |
|--------------------|-----------------|---------------|
| Classes 44 and 45  | $6,600,000      | $2,200,000    |

#### Residential Performance

|          | Dates               | Target | Applied | Graduated | Received GED/HiSET | Received HS Credits | Received HS Diploma |
|----------|---------------------|--------|---------|-----------|--------------------|---------------------|---------------------|
| Class 44 | Jan. 2015–June 2015 | 300    | 476     | 195       | 113                | ~                   | ~                   |
| Class 45 | July 2015–Dec. 2015 | 300    | 420     | 174       | 85                 | ~                   | ~                   |

#### Physical Fitness

|          | Curl-Ups | | Push-Ups | | 1-Mile Run | |
|----------|----------|-------|----------|-------|------------|-------|
|          | Initial  | Final | Initial  | Final | Initial    | Final |
| Class 44 | 23.4     | 49.3  | *        | *     | 10:17      | 09:08 |
| Class 45 | 29.8     | 57.1  | 19.3     | 40.7  | 10:26      | 08:42 |

#### Responsible Citizenship

|          | Voting | | Selective Service | |
|----------|----------|------------|----------|------------|
|          | Eligible | Registered | Eligible | Registered |
| Class 44 | 52       | 52         | 34       | 34         |
| Class 45 | 38       | 38         | 27       | 27         |

#### Service to Community

|          | Hours of Service | Dollar Value/Hr | Total Value |
|----------|------------------|-----------------|-------------|
| Class 44 | 11,896           | $25.34          | $301,445    |
| Class 45 | 11,854           | $25.34          | $300,380    |

#### Postresidential Performance Status

|          | Graduated | Contacted | Placed | Education | Employment | Military | Other |
|----------|-----------|-----------|--------|-----------|------------|----------|-------|
| Class 44 |           |           |        |           |            |          |       |
| Month 1  | 195       | 108       | 59     | 27        | 46         | 0        | 3     |
| Month 6  | 195       | 97        | 104    | 60        | 60         | 1        | 1     |
| Month 12 | 195       | 101       | 71     | 60        | 68         | 1        | 0     |
| Class 45 |           |           |        |           |            |          |       |
| Month 1  | 174       | 99        | 53     | 30        | 24         | 0        | 7     |
| Month 6  | 174       | 72        | 46     | 22        | 23         | 5        | 2     |

* Did not report
~ Does not award

**Table C.13**
**Indiana Profile**

| HOOSIER YOUTH CHALLENGE ACADEMY, ESTABLISHED 2007 | |
|---|---|

| Graduates since inception: 1,253 | Program type: GED or HiSET |
|---|---|

### Staffing

| | Instructional | Cadre | Administrative | Other | Total |
|---|---|---|---|---|---|
| Full-time | 4 | 28 | 14 | 2 | 48 |
| Part-time | 0 | 0 | 0 | 0 | 0 |

### Funding

| | Federal Funding | State Funding |
|---|---|---|
| Classes 44 and 45 | $3,195,000 | $1,064,000 |

### Residential Performance

| | Dates | Target | Applied | Graduated | Received GED/ HiSET | Received HS Credits | Received HS Diploma |
|---|---|---|---|---|---|---|---|
| Class 44 | Jan. 2015–June 2015 | 100 | 130 | 69 | 36 | ~ | ~ |
| Class 45 | July 2015–Dec. 2015 | 100 | 172 | 80 | 51 | ~ | ~ |

### Physical Fitness

| | Curl-Ups | | Push-Ups | | 1-Mile Run | |
|---|---|---|---|---|---|---|
| | Initial | Final | Initial | Final | Initial | Final |
| Class 44 | 29.8 | 47.5 | * | * | 17:30 | 08:10 |
| Class 45 | 30.3 | 52.8 | * | * | 10:19 | 08:16 |

### Responsible Citizenship

| | Voting | | Selective Service | |
|---|---|---|---|---|
| | Eligible | Registered | Eligible | Registered |
| Class 44 | 11 | 11 | 29 | 29 |
| Class 45 | 15 | 15 | 32 | 32 |

### Service to Community

| | Hours of Service | Dollar Value/Hr | Total Value |
|---|---|---|---|
| Class 44 | 4,059 | $22.69 | $92,087 |
| Class 45 | 4,848 | $22.69 | $110,001 |

### Postresidential Performance Status

| | Graduated | Contacted | Placed | Education | Employment | Military | Other |
|---|---|---|---|---|---|---|---|
| Class 44 | | | | | | | |
| Month 1 | 69 | 69 | 18 | 5 | 9 | 0 | 4 |
| Month 6 | 69 | 69 | 28 | 18 | 7 | 2 | 1 |
| Month 12 | 69 | 69 | 35 | 15 | 11 | 8 | 1 |
| Class 45 | | | | | | | |
| Month 1 | 80 | 80 | 8 | 2 | 4 | 0 | 2 |
| Month 6 | 80 | 80 | 22 | 5 | 15 | 1 | 1 |

* Did not report

~ Does not award

**Table C.14**
**Kentucky, Bluegrass ChalleNGe Academy Profile**

| BLUEGRASS CHALLENGE ACADEMY, ESTABLISHED 1999 |
| :---: |

| Graduates since inception: 2,806 | Program type: Credit Recovery, GED or HiSET |

**Staffing**

|  | Instructional | Cadre | Administrative | Other | Total |
| --- | --- | --- | --- | --- | --- |
| Full-time | 6 | 23 | 12 | 3 | 44 |
| Part-time | 0 | 5 | 1 | 1 | 7 |

**Funding**

|  | Federal Funding | State Funding |
| --- | --- | --- |
| Classes 44 and 45 | $2,555,160 | $851,720 |

**Residential Performance**

|  | Dates | Target | Applied | Graduated | Received GED/ HiSET | Received HS Credits | Received HS Diploma |
| --- | --- | --- | --- | --- | --- | --- | --- |
| Class 44 | Jan. 2015–June 2015 | 100 | 136 | 94 | 14 | 93 | ~ |
| Class 45 | July 2015–Dec. 2015 | 100 | 118 | 50 | 38 | 38 | ~ |

**Physical Fitness**

|  | Curl-Ups | | Push-Ups | | 1-Mile Run | |
| --- | --- | --- | --- | --- | --- | --- |
|  | Initial | Final | Initial | Final | Initial | Final |
| Class 44 | 29.6 | 37.9 | 22.0 | 32.8 | 12:14 | 10:40 |
| Class 45 | 22.1 | 41.1 | 18.7 | 37.1 | 12:21 | 10:18 |

**Responsible Citizenship**

|  | Voting | | Selective Service | |
| --- | --- | --- | --- | --- |
|  | Eligible | Registered | Eligible | Registered |
| Class 44 | 6 | 6 | 12 | 12 |
| Class 45 | 7 | 7 | 13 | 13 |

**Service to Community**

|  | Hours of Service | Dollar Value/Hr | Total Value |
| --- | --- | --- | --- |
| Class 44 | 3,140 | $21.16 | $66,432 |
| Class 45 | 3,940 | $21.16 | $83,370 |

**Postresidential Performance Status**

|  | Graduated | Contacted | Placed | Education | Employment | Military | Other |
| --- | --- | --- | --- | --- | --- | --- | --- |
| Class 44 |  |  |  |  |  |  |  |
| Month 1 | 94 | 94 | 92 | 90 | 0 | 2 | 0 |
| Month 6 | 94 | 94 | 85 | 83 | 0 | 2 | 0 |
| Month 12 | 94 | 94 | 79 | 77 | 0 | 2 | 0 |
| Class 45 |  |  |  |  |  |  |  |
| Month 1 | 50 | 50 | 47 | 46 | 1 | 0 | 0 |
| Month 6 | 50 | 50 | 46 | 45 | 1 | 0 | 0 |

~ Does not award

**Table C.15**
**Kentucky, Appalachian ChalleNGe Program Profile**

| APPALACHIAN CHALLENGE PROGRAM, ESTABLISHED 2012 | |
|---|---|
| Graduates since inception: 576 | Program type: Credit Recovery, GED or HiSET |

**Staffing**

| | Instructional | Cadre | Administrative | Other | Total |
|---|---|---|---|---|---|
| Full-time | 5 | 22 | 17 | 4 | 48 |
| Part-time | 0 | 2 | 0 | 0 | 2 |

**Funding**

| | Federal Funding | State Funding |
|---|---|---|
| Classes 44 and 45 | $2,647,000 | $882,333 |

**Residential Performance**

| | Dates | Target | Applied | Graduated | Received GED/ HiSET | Received HS Credits | Received HS Diploma |
|---|---|---|---|---|---|---|---|
| Class 44 | Jan. 2015–June 2015 | 100 | 118 | 84 | 16 | 66 | ~ |
| Class 45 | July 2015–Dec. 2015 | 100 | 187 | 110 | 5 | 105 | ~ |

**Physical Fitness**

| | Curl-Ups | | Push-Ups | | 1-Mile Run | |
|---|---|---|---|---|---|---|
| | Initial | Final | Initial | Final | Initial | Final |
| Class 44 | 32.7 | 55.5 | 37.3 | 53.5 | 10:58 | 08:36 |
| Class 45 | 36.0 | 60.4 | 23.0 | 53.8 | 09:35 | 08:22 |

**Responsible Citizenship**

| | Voting | | Selective Service | |
|---|---|---|---|---|
| | Eligible | Registered | Eligible | Registered |
| Class 44 | 15 | 15 | 15 | 15 |
| Class 45 | 20 | 20 | 20 | 20 |

**Service to Community**

| | Hours of Service | Dollar Value/Hr | Total Value |
|---|---|---|---|
| Class 44 | 5,761 | $21.16 | $121,903 |
| Class 45 | 7,270 | $21.16 | $153,833 |

**Postresidential Performance Status**

| | Graduated | Contacted | Placed | Education | Employment | Military | Other |
|---|---|---|---|---|---|---|---|
| Class 44 | | | | | | | |
| Month 1 | 84 | 84 | 60 | 55 | 5 | 0 | 0 |
| Month 6 | 84 | 84 | 68 | 61 | 8 | 0 | 3 |
| Month 12 | 84 | 84 | 61 | 44 | 19 | 1 | 4 |
| Class 45 | | | | | | | |
| Month 1 | 110 | 106 | 94 | 81 | 15 | 0 | 0 |
| Month 6 | 110 | 106 | 94 | 74 | 21 | 2 | 0 |

~ Does not award

**Table C.16**
**Louisiana, Camp Beauregard Profile**

| LOUISIANA YOUTH CHALLENGE PROGRAM—CAMP BEAUREGARD, ESTABLISHED 1993 |
|---|

Graduates since inception: 9,159                          Program type: GED or HiSET

### Staffing

| | Instructional | Cadre | Administrative | Other | Total |
|---|---|---|---|---|---|
| Full-time | 15 | 52 | 14 | 32 | 113 |
| Part-time | 0 | 6 | 3 | 3 | 12 |

### Funding

| | Federal Funding | State Funding |
|---|---|---|
| Classes 44 and 45 | $6,375,000 | $2,125,000 |

### Residential Performance

| | Dates | Target | Applied | Graduated | Received GED/ HiSET | Received HS Credits | Received HS Diploma |
|---|---|---|---|---|---|---|---|
| Class 44 | Jan. 2015–June 2015 | 250 | 498 | 250 | 58 | ~ | ~ |
| Class 45 | July 2015–Dec. 2015 | 250 | 567 | 221 | 78 | ~ | ~ |

### Physical Fitness

| | Curl-Ups | | Push-Ups | | 1-Mile Run | |
|---|---|---|---|---|---|---|
| | Initial | Final | Initial | Final | Initial | Final |
| Class 44 | 31.8 | 51.3 | 30.3 | 41.0 | 09:09 | 08:08 |
| Class 45 | 34.5 | 36.9 | 29.8 | 47.1 | 09:30 | 07:14 |

### Responsible Citizenship

| | Voting | | Selective Service | |
|---|---|---|---|---|
| | Eligible | Registered | Eligible | Registered |
| Class 44 | 32 | 30 | 121 | 121 |
| Class 45 | 35 | 32 | 108 | 108 |

### Service to Community

| | Hours of Service | Dollar Value/Hr | Total Value |
|---|---|---|---|
| Class 44 | 11,888 | $22.67 | $269,501 |
| Class 45 | 10,145 | $22.67 | $229,987 |

### Postresidential Performance Status

| | Graduated | Contacted | Placed | Education | Employment | Military | Other |
|---|---|---|---|---|---|---|---|
| Class 44 | | | | | | | |
| Month 1 | 250 | 249 | 228 | 41 | 153 | 1 | 33 |
| Month 6 | 250 | 240 | 210 | 97 | 106 | 3 | 4 |
| Month 12 | 250 | 236 | 232 | 70 | 124 | 6 | 32 |
| Class 45 | | | | | | | |
| Month 1 | 221 | 217 | 189 | 44 | 116 | 0 | 29 |
| Month 6 | 221 | 212 | 178 | 62 | 102 | 6 | 8 |

~ Does not award

**Table C.17**
**Louisiana, Camp Minden Profile**

| LOUISIANA YOUTH CHALLENGE PROGRAM—CAMP MINDEN, ESTABLISHED 2002 |
|---|

| Graduates since inception: 4,219 | Program type: GED or HiSET, Other |
|---|---|

### Staffing

| | Instructional | Cadre | Administrative | Other | Total |
|---|---|---|---|---|---|
| Full-time | 13 | 43 | 32 | 0 | 88 |
| Part-time | 0 | 0 | 1 | 0 | 1 |

### Funding

| | Federal Funding | State Funding |
|---|---|---|
| Classes 44 and 45 | $5,100,000 | $1,700,000 |

### Residential Performance

| | Dates | Target | Applied | Graduated | Received GED/ HiSET | Received HS Credits | Received HS Diploma |
|---|---|---|---|---|---|---|---|
| Class 44 | Jan. 2015–June 2015 | 200 | 445 | 220 | 101 | ~ | ~ |
| Class 45 | July 2015–Dec. 2015 | 200 | 366 | 200 | 82 | ~ | ~ |

### Physical Fitness

| | Curl-Ups | | Push-Ups | | 1-Mile Run | |
|---|---|---|---|---|---|---|
| | Initial | Final | Initial | Final | Initial | Final |
| Class 44 | 29.6 | 36.1 | 25.3 | 42.9 | 09:03 | 07:27 |
| Class 45 | * | * | * | * | * | * |

### Responsible Citizenship

| | Voting | | Selective Service | |
|---|---|---|---|---|
| | Eligible | Registered | Eligible | Registered |
| Class 44 | 41 | 40 | 95 | 94 |
| Class 45 | 30 | 30 | 54 | 44 |

### Service to Community

| | Hours of Service | Dollar Value/Hr | Total Value |
|---|---|---|---|
| Class 44 | 9,352 | $22.67 | $212,010 |
| Class 45 | 10,187 | $22.67 | $230,939 |

### Postresidential Performance Status

| | Graduated | Contacted | Placed | Education | Employment | Military | Other |
|---|---|---|---|---|---|---|---|
| Class 44 | | | | | | | |
| Month 1 | 220 | 219 | 186 | 57 | 142 | 5 | 3 |
| Month 6 | 220 | 219 | 181 | 55 | 131 | 6 | 4 |
| Month 12 | 220 | 218 | 188 | 51 | 145 | 5 | 6 |
| Class 45 | | | | | | | |
| Month 1 | 200 | 200 | 174 | 55 | 129 | 2 | 3 |
| Month 6 | 200 | 199 | 173 | 55 | 124 | 2 | 3 |

* Did not report

~ Does not award

**Table C.18**
**Louisiana, Gillis Long Profile**

| LOUISIANA YOUTH CHALLENGE PROGRAM—GILLIS LONG, ESTABLISHED 1999 |
|---|

Graduates since inception: 7,204      Program type: GED or HiSET

### Staffing

|  | Instructional | Cadre | Administrative | Other | Total |
|---|---|---|---|---|---|
| Full-time | 13 | 50 | 13 | 45 | 121 |
| Part-time | 2 | 1 | 0 | 4 | 7 |

### Funding

|  | Federal Funding | State Funding |
|---|---|---|
| Classes 44 and 45 | $6,375,000 | $2,125,000 |

### Residential Performance

|  | Dates | Target | Applied | Graduated | Received GED/ HiSET | Received HS Credits | Received HS Diploma |
|---|---|---|---|---|---|---|---|
| Class 44 | Apr. 2015–Sept. 2015 | 250 | 468 | 263 | 102 | ~ | 1 |
| Class 45 | Oct. 2015–Mar. 2016 | 250 | 458 | 250 | 95 | ~ | ~ |

### Physical Fitness

|  | Curl-Ups | | Push-Ups | | 1-Mile Run | |
|---|---|---|---|---|---|---|
|  | Initial | Final | Initial | Final | Initial | Final |
| Class 44 | 26.4 | 37.9 | 20.7 | 29.4 | 09:17 | 10:52 |
| Class 45 | 23.9 | 43.0 | 20.5 | 41.0 | 12:04 | 10:30 |

### Responsible Citizenship

|  | Voting | | Selective Service | |
|---|---|---|---|---|
|  | Eligible | Registered | Eligible | Registered |
| Class 44 | 28 | 28 | 23 | 23 |
| Class 45 | 49 | 49 | 34 | 34 |

### Service to Community

|  | Hours of Service | Dollar Value/Hr | Total Value |
|---|---|---|---|
| Class 44 | 16,411 | $22.67 | $372,037 |
| Class 45 | 19,807 | $22.67 | $449,025 |

### Postresidential Performance Status

|  | Graduated | Contacted | Placed | Education | Employment | Military | Other |
|---|---|---|---|---|---|---|---|
| Class 44 |  |  |  |  |  |  |  |
| Month 1 | 263 | 263 | 221 | 80 | 123 | 1 | 42 |
| Month 6 | 263 | 263 | 222 | 103 | 133 | 3 | 26 |
| Month 12 | 263 | 263 | 218 | 102 | 143 | 8 | 26 |
| Class 45 |  |  |  |  |  |  |  |
| Month 1 | 250 | 250 | 198 | 44 | 106 | 3 | 20 |
| Month 6 | 250 | 250 | 194 | 51 | 126 | 4 | 7 |

~ Does not award

**Table C.19**
**Maryland Profile**

### FREESTATE CHALLENGE ACADEMY, ESTABLISHED 1993

Graduates since inception: 3,914                    Program type: GED or HiSET

**Staffing**

|  | Instructional | Cadre | Administrative | Other | Total |
|---|---|---|---|---|---|
| Full-time | 4 | 31 | 11 | 12 | 58 |
| Part-time | 0 | 2 | 2 | 0 | 4 |

**Funding**

|  | Federal Funding | State Funding |
|---|---|---|
| Classes 44 and 45 | $2,334,400 | $776,467 |

**Residential Performance**

|  | Dates | Target | Applied | Graduated | Received GED/ HiSET | Received HS Credits | Received HS Diploma |
|---|---|---|---|---|---|---|---|
| Class 44 | Jan. 2015–June 2015 | 84 | 189 | 84 | 34 | ~ | 27 |
| Class 45 | July 2015–Dec. 2015 | 107 | 255 | 107 | 62 | ~ | 62 |

**Physical Fitness**

|  | Curl-Ups | | Push-Ups | | 1-Mile Run | |
|---|---|---|---|---|---|---|
|  | Initial | Final | Initial | Final | Initial | Final |
| Class 44 | 25.1 | 48.3 | 24.3 | 41.6 | 10:45 | 09:01 |
| Class 45 | 31.5 | 58.3 | 25.5 | 46.9 | 12:09 | 08:39 |

**Responsible Citizenship**

|  | Voting | | Selective Service | |
|---|---|---|---|---|
|  | Eligible | Registered | Eligible | Registered |
| Class 44 | 19 | 19 | 32 | 32 |
| Class 45 | 42 | 42 | 46 | 46 |

**Service to Community**

|  | Hours of Service | Dollar Value/Hr | Total Value |
|---|---|---|---|
| Class 44 | 4,049 | $26.64 | $107,865 |
| Class 45 | 4,855 | $26.64 | $129,326 |

**Postresidential Performance Status**

|  | Graduated | Contacted | Placed | Education | Employment | Military | Other |
|---|---|---|---|---|---|---|---|
| Class 44 |  |  |  |  |  |  |  |
| Month 1 | 84 | 84 | 19 | 9 | 8 | 1 | 1 |
| Month 6 | 84 | 83 | 57 | 20 | 36 | 1 | 0 |
| Month 12 | 84 | 82 | 54 | 19 | 33 | 1 | 1 |
| Class 45 |  |  |  |  |  |  |  |
| Month 1 | 107 | 107 | 27 | 6 | 19 | 0 | 2 |
| Month 6 | 107 | 106 | 60 | 15 | 43 | 0 | 2 |

~ Does not award

**Table C.20**
**Michigan Profile**

### MICHIGAN YOUTH CHALLENGE ACADEMY, ESTABLISHED 1999

Graduates since inception: 3,132

Program type: Credit Recovery, High School Diploma, GED or HiSET

#### Staffing

|  | Instructional | Cadre | Administrative | Other | Total |
|---|---|---|---|---|---|
| Full-time | 8 | 19 | 16 | 0 | 43 |
| Part-time | 0 | 0 | 0 | 0 | 0 |

#### Funding

|  | Federal Funding | State Funding |
|---|---|---|
| Classes 44 and 45 | $3,000,000 | $1,000,000 |

#### Residential Performance

|  | Dates | Target | Applied | Graduated | Received GED/ HiSET | Received HS Credits | Received HS Diploma |
|---|---|---|---|---|---|---|---|
| Class 44 | Jan. 2015–June 2015 | 114 | 191 | 106 | * | 106 | * |
| Class 45 | July 2015–Dec. 2015 | 114 | 197 | 107 | * | 107 | * |

#### Physical Fitness

|  | Curl-Ups | | Push-Ups | | 1-Mile Run | |
|---|---|---|---|---|---|---|
|  | Initial | Final | Initial | Final | Initial | Final |
| Class 44 | 41.1 | 53.8 | 32.1 | 56.7 | 08:27 | 07:48 |
| Class 45 | 41.5 | 50.8 | 37.3 | 54.2 | 09:27 | 07:36 |

#### Responsible Citizenship

|  | Voting | | Selective Service | |
|---|---|---|---|---|
|  | Eligible | Registered | Eligible | Registered |
| Class 44 | 24 | 0 | 27 | 0 |
| Class 45 | 33 | 0 | 33 | 9 |

#### Service to Community

|  | Hours of Service | Dollar Value/Hr | Total Value |
|---|---|---|---|
| Class 44 | 4,869 | $23.54 | $114,616 |
| Class 45 | 3,992 | $23.54 | $93,960 |

#### Postresidential Performance Status

|  | Graduated | Contacted | Placed | Education | Employment | Military | Other |
|---|---|---|---|---|---|---|---|
| Class 44 |  |  |  |  |  |  |  |
| Month 1 | 106 | * | 41 | 17 | 44 | 8 | 4 |
| Month 6 | 106 | * | 73 | 50 | 40 | 12 | 1 |
| Month 12 | 106 | * | 70 | 36 | 49 | 14 | 2 |
| Class 45 |  |  |  |  |  |  |  |
| Month 1 | 107 | * | 67 | 65 | 12 | 9 | 1 |
| Month 6 | 107 | * | 40 | 25 | 31 | 7 | 0 |

* Did not report

**Table C.21**
**Mississippi Profile**

### MISSISSIPPI YOUTH CHALLENGE ACADEMY, ESTABLISHED 1994

| Graduates since inception: 8,422 | | | Program type: High School Diploma | | | |

#### Staffing

| | Instructional | Cadre | Administrative | Other | Total |
|---|---|---|---|---|---|
| Full-time | 9 | 45 | 19 | 25 | 98 |
| Part-time | 6 | 2 | 0 | 0 | 8 |

#### Funding

| | Federal Funding | State Funding |
|---|---|---|
| Classes 44 and 45 | $4,200,000 | $1,400,000 |

#### Residential Performance

| | Dates | Target | Applied | Graduated | Received GED/HiSET | Received HS Credits | Received HS Diploma |
|---|---|---|---|---|---|---|---|
| Class 44 | Jan. 2015–June 2015 | 200 | 408 | 168 | 94 | ~ | 94 |
| Class 45 | July 2015–Dec. 2015 | 200 | 520 | 206 | 129 | ~ | 129 |

#### Physical Fitness

| | Curl-Ups | | Push-Ups | | 1-Mile Run | |
|---|---|---|---|---|---|---|
| | Initial | Final | Initial | Final | Initial | Final |
| Class 44 | 28.4 | 47.8 | 22.3 | 42.0 | 11:46 | 08:31 |
| Class 45 | 35.3 | 53.2 | 21.8 | 46.3 | 10:51 | 07:55 |

#### Responsible Citizenship

| | Voting | | Selective Service | |
|---|---|---|---|---|
| | Eligible | Registered | Eligible | Registered |
| Class 44 | 40 | 40 | 57 | 57 |
| Class 45 | 51 | 51 | 81 | 81 |

#### Service to Community

| | Hours of Service | Dollar Value/Hr | Total Value |
|---|---|---|---|
| Class 44 | 14,490 | $19.51 | $282,690 |
| Class 45 | 13,247 | $19.51 | $258,439 |

#### Postresidential Performance Status

| | Graduated | Contacted | Placed | Education | Employment | Military | Other |
|---|---|---|---|---|---|---|---|
| Class 44 | | | | | | | |
| Month 1 | 168 | 149 | 96 | 48 | 62 | 8 | 24 |
| Month 6 | 168 | 127 | 93 | 65 | 82 | 10 | 21 |
| Month 12 | 168 | 113 | 79 | 49 | 87 | 14 | 19 |
| Class 45 | | | | | | | |
| Month 1 | 206 | 192 | 138 | 64 | 72 | 4 | 29 |
| Month 6 | 206 | 157 | 126 | 68 | 102 | 7 | 27 |

~ Does not award

**Table C.22**
**Montana Profile**

| MONTANA YOUTH CHALLENGE ACADEMY, ESTABLISHED 1999 |
|---|

| Graduates since inception: 2,453 | Program type: GED or HiSET, Credit Recovery |
|---|---|

### Staffing

| | Instructional | Cadre | Administrative | Other | Total |
|---|---|---|---|---|---|
| Full-time | 5 | 23 | 8 | 11 | 47 |
| Part-time | 0 | 5 | 0 | 1 | 6 |

### Funding

| | Federal Funding | State Funding |
|---|---|---|
| Classes 44 and 45 | $3,393,500 | $1,131,166 |

### Residential Performance

| | Dates | Target | Applied | Graduated | Received GED/ HiSET | Received HS Credits | Received HS Diploma |
|---|---|---|---|---|---|---|---|
| Class 44 | Jan. 2015–June 2015 | 100 | 126 | 84 | 42 | 84 | ~ |
| Class 45 | July 2015–Dec. 2015 | 100 | 120 | 74 | 34 | 74 | ~ |

### Physical Fitness

| | Curl-Ups | | Push-Ups | | 1-Mile Run | |
|---|---|---|---|---|---|---|
| | Initial | Final | Initial | Final | Initial | Final |
| Class 44 | 35.7 | 49.9 | * | * | 10:35 | 08:35 |
| Class 45 | 40.9 | 48.4 | * | * | 09:59 | 08:04 |

### Responsible Citizenship

| | Voting | | Selective Service | |
|---|---|---|---|---|
| | Eligible | Registered | Eligible | Registered |
| Class 44 | 19 | 0 | 33 | 33 |
| Class 45 | 15 | 0 | 25 | 25 |

### Service to Community

| | Hours of Service | Dollar Value/Hr | Total Value |
|---|---|---|---|
| Class 44 | 4,270 | $20.44 | $87,284 |
| Class 45 | 4,338 | $20.44 | $88,677 |

### Postresidential Performance Status

| | Graduated | Contacted | Placed | Education | Employment | Military | Other |
|---|---|---|---|---|---|---|---|
| Class 44 | | | | | | | |
| Month 1 | 84 | 79 | 65 | 11 | 42 | 2 | 10 |
| Month 6 | 84 | 71 | 52 | 23 | 19 | 2 | 8 |
| Month 12 | 84 | 73 | 60 | 17 | 34 | 3 | 6 |
| Class 45 | | | | | | | |
| Month 1 | 74 | 74 | 61 | 43 | 10 | 2 | 6 |
| Month 6 | 74 | 70 | 57 | 20 | 29 | 3 | 5 |

* Did not report

~ Does not award

**Table C.23**
**North Carolina, New London Profile**

| TARHEEL CHALLENGE ACADEMY—NEW LONDON, ESTABLISHED 2015 |
|---|

| Graduates since inception: 50 | Program type: GED or HiSET |
|---|---|

### Staffing

| | Instructional | Cadre | Administrative | Other | Total |
|---|---|---|---|---|---|
| Full-time | 7 | 21 | 13 | 11 | 52 |
| Part-time | 0 | 0 | 0 | 0 | 0 |

### Funding

| | Federal Funding | State Funding |
|---|---|---|
| Classes 44 and 45 | $2,129,280 | $708,797 |

### Residential Performance

| | Dates | Target | Applied | Graduated | Received GED/ HiSET | Received HS Credits | Received HS Diploma |
|---|---|---|---|---|---|---|---|
| Class 44 | ^ | ^ | ^ | ^ | ^ | ^ | ^ |
| Class 45 | Nov. 2015–Apr. 2016 | 100 | 104 | 50 | 27 | ~ | ~ |

### Physical Fitness

| | Curl-Ups | | Push-Ups | | 1-Mile Run | |
|---|---|---|---|---|---|---|
| | Initial | Final | Initial | Final | Initial | Final |
| Class 44 | ^ | ^ | ^ | ^ | ^ | ^ |
| Class 45 | 35.4 | 52.1 | 24.2 | 39.6 | 07:33 | 06:38 |

### Responsible Citizenship

| | Voting | | Selective Service | |
|---|---|---|---|---|
| | Eligible | Registered | Eligible | Registered |
| Class 44 | ^ | ^ | ^ | ^ |
| Class 45 | 15 | 15 | 12 | 12 |

### Service to Community

| | Hours of Service | Dollar Value/Hr | Total Value |
|---|---|---|---|
| Class 44 | ^ | ^ | ^ |
| Class 45 | 2,000 | $21.88 | $43,760 |

### Postresidential Performance Status

| | Graduated | Contacted | Placed | Education | Employment | Military | Other |
|---|---|---|---|---|---|---|---|
| Class 44 | | | | | | | |
| Month 1 | ^ | ^ | ^ | ^ | ^ | ^ | ^ |
| Month 6 | ^ | ^ | ^ | ^ | ^ | ^ | ^ |
| Month 12 | ^ | ^ | ^ | ^ | ^ | ^ | ^ |
| Class 45 | | | | | | | |
| Month 1 | 50 | 50 | 30 | 3 | 22 | 1 | 4 |
| Month 6 | 50 | 50 | 33 | 7 | 24 | 2 | 0 |

^ Newly operational

~ Does not award

**Table C.24**
**North Carolina, Salemburg Profile**

| TARHEEL CHALLENGE ACADEMY—SALEMBURG, ESTABLISHED 1994 |
|---|

**Graduates since inception: 4,320**                          **Program type: GED or HiSET**

### Staffing

|  | Instructional | Cadre | Administrative | Other | Total |
|---|---|---|---|---|---|
| Full-time | 7 | 25 | 27 | 0 | 59 |
| Part-time | 2 | 1 | 0 | 0 | 3 |

### Funding

|  | Federal Funding | State Funding |
|---|---|---|
| Classes 44 and 45 | $3,000,000 | $1,000,000 |

### Residential Performance

|  | Dates | Target | Applied | Graduated | Received GED/ HiSET | Received HS Credits | Received HS Diploma |
|---|---|---|---|---|---|---|---|
| Class 44 | Jan. 2015–June 2015 | 125 | 365 | 106 | 48 | ~ | ~ |
| Class 45 | July 2015–Dec. 2015 | 125 | 455 | 140 | 47 | ~ | ~ |

### Physical Fitness

|  | Curl-Ups | | Push-Ups | | 1-Mile Run | |
|---|---|---|---|---|---|---|
|  | Initial | Final | Initial | Final | Initial | Final |
| Class 44 | 28.5 | 45.5 | 21.0 | 38.5 | 11:23 | 08:20 |
| Class 45 | 30.4 | 42.5 | 21.9 | 36.0 | 10:27 | 07:50 |

### Responsible Citizenship

|  | Voting | | Selective Service | |
|---|---|---|---|---|
|  | Eligible | Registered | Eligible | Registered |
| Class 44 | 23 | 23 | 18 | 18 |
| Class 45 | 33 | 33 | 30 | 30 |

### Service to Community

|  | Hours of Service | Dollar Value/Hr | Total Value |
|---|---|---|---|
| Class 44 | 8,303 | $21.88 | $181,677 |
| Class 45 | 9,765 | $21.88 | $213,658 |

### Postresidential Performance Status

|  | Graduated | Contacted | Placed | Education | Employment | Military | Other |
|---|---|---|---|---|---|---|---|
| Class 44 |  |  |  |  |  |  |  |
| Month 1 | 106 | 106 | 15 | 3 | 12 | 0 | 0 |
| Month 6 | 106 | 106 | 32 | 19 | 9 | 1 | 3 |
| Month 12 | 106 | 106 | 25 | 17 | 6 | 1 | 1 |
| Class 45 |  |  |  |  |  |  |  |
| Month 1 | 140 | 140 | 36 | 9 | 16 | 0 | 11 |
| Month 6 | 140 | 140 | 58 | 26 | 23 | 1 | 8 |

~ Does not award

**Table C.25**
**New Jersey Profile**

### NEW JERSEY YOUTH CHALLENGE ACADEMY, ESTABLISHED 1994

| Graduates since inception: 3,626 | | | | | | Program type: GED or HiSET | | |
|---|---|---|---|---|---|---|---|---|

**Staffing**

| | Instructional | Cadre | Administrative | Other | Total |
|---|---|---|---|---|---|
| Full-time | 4 | 16 | 13 | 0 | 33 |
| Part-time | 3 | 14 | 5 | 0 | 22 |

**Funding**

| | Federal Funding | State Funding |
|---|---|---|
| Classes 44 and 45 | $2,700,000 | $1,227,861 |

**Residential Performance**

| | Dates | Target | Applied | Graduated | Received GED/ HiSET | Received HS Credits | Received HS Diploma |
|---|---|---|---|---|---|---|---|
| Class 44 | Oct. 2014–Mar. 2015 | 100 | 291 | 100 | 44 | ~ | 44 |
| Class 45 | Apr. 2015–Sept. 2015 | 100 | 290 | 99 | 44 | ~ | 44 |

**Physical Fitness**

| | Curl-Ups | | Push-Ups | | 1-Mile Run | |
|---|---|---|---|---|---|---|
| | Initial | Final | Initial | Final | Initial | Final |
| Class 44 | 42.7 | 55.2 | 32.9 | 52.6 | 10:05 | 07:59 |
| Class 45 | 36.4 | 44.9 | 29.1 | 45.7 | 07:32 | 06:46 |

**Responsible Citizenship**

| | Voting | | Selective Service | |
|---|---|---|---|---|
| | Eligible | Registered | Eligible | Registered |
| Class 44 | 38 | 38 | 33 | 33 |
| Class 45 | 29 | 29 | 21 | 21 |

**Service to Community**

| | Hours of Service | Dollar Value/Hr | Total Value |
|---|---|---|---|
| Class 44 | 4,350 | $26.70 | $116,145 |
| Class 45 | 5,473 | $26.70 | $146,116 |

**Postresidential Performance Status**

| | Graduated | Contacted | Placed | Education | Employment | Military | Other |
|---|---|---|---|---|---|---|---|
| Class 44 | | | | | | | |
| Month 1 | 100 | 100 | 29 | 12 | 15 | 2 | 4 |
| Month 6 | 100 | 100 | 89 | 36 | 55 | 10 | 7 |
| Month 12 | 100 | 100 | 92 | 36 | 58 | 14 | 6 |
| Class 45 | | | | | | | |
| Month 1 | 99 | 100 | 36 | 14 | 21 | 0 | 5 |
| Month 6 | 99 | 100 | 79 | 33 | 64 | 4 | 5 |

~ Does not award

**Table C.26**
**New Mexico Profile**

| NEW MEXICO YOUTH CHALLENGE ACADEMY, ESTABLISHED 2001 |
|---|

| Graduates since inception: 2,159 | Program type: GED |
|---|---|

**Staffing**

|  | Instructional | Cadre | Administrative | Other | Total |
|---|---|---|---|---|---|
| Full-time | 15 | 22 | 18 | 0 | 55 |
| Part-time | 0 | 0 | 0 | 0 | 0 |

**Funding**

|  | Federal Funding | State Funding |
|---|---|---|
| Classes 44 and 45 | $2,400,000 | $800,000 |

**Residential Performance**

|  | Dates | Target | Applied | Graduated | Received GED/ HiSET | Received HS Credits | Received HS Diploma |
|---|---|---|---|---|---|---|---|
| Class 44 | Jan. 2015–June 2015 | 100 | 143 | 80 | 31 | ~ | ~ |
| Class 45 | July 2015–Dec. 2015 | 100 | 162 | 94 | 77 | ~ | ~ |

**Physical Fitness**

|  | Curl-Ups | | Push-Ups | | 1-Mile Run | |
|---|---|---|---|---|---|---|
|  | Initial | Final | Initial | Final | Initial | Final |
| Class 44 | 31.7 | 50.9 | 34.3 | 64.1 | 08:04 | 06:18 |
| Class 45 | 35.2 | 53.3 | 34.9 | 59.0 | 08:52 | 06:18 |

**Responsible Citizenship**

|  | Voting | | Selective Service | |
|---|---|---|---|---|
|  | Eligible | Registered | Eligible | Registered |
| Class 44 | 91 | 52 | 63 | 29 |
| Class 45 | 73 | 19 | 42 | 34 |

**Service to Community**

|  | Hours of Service | Dollar Value/Hr | Total Value |
|---|---|---|---|
| Class 44 | 6,491 | $19.91 | $129,226 |
| Class 45 | 4,322 | $19.91 | $86,051 |

**Postresidential Performance Status**

|  | Graduated | Contacted | Placed | Education | Employment | Military | Other |
|---|---|---|---|---|---|---|---|
| Class 44 |  |  |  |  |  |  |  |
| Month 1 | 80 | 80 | 73 | 8 | 40 | 2 | 40 |
| Month 6 | 80 | 80 | 72 | 25 | 51 | 5 | 12 |
| Month 12 | 80 | 80 | 70 | 9 | 58 | 4 | 15 |
| Class 45 |  |  |  |  |  |  |  |
| Month 1 | 94 | 92 | 80 | 30 | 51 | 2 | 27 |
| Month 6 | 94 | 92 | 79 | 10 | 60 | 2 | 15 |

~ Does not award

**Table C.27**
**Oklahoma Profile**

### THUNDERBIRD YOUTH ACADEMY, ESTABLISHED 1993

Graduates since inception: 4,225                    Program type: Credit Recovery, GED

#### Staffing

|  | Instructional | Cadre | Administrative | Other | Total |
|---|---|---|---|---|---|
| Full-time | 7 | 38 | 21 | 8 | 74 |
| Part-time | 0 | 0 | 0 | 0 | 0 |

#### Funding

|  | Federal Funding | State Funding |
|---|---|---|
| Classes 44 and 45 | $2,805,000 | $935,000 |

#### Residential Performance

|  | Dates | Target | Applied | Graduated | Received GED/HiSET | Received HS Credits | Received HS Diploma |
|---|---|---|---|---|---|---|---|
| Class 44 | Jan. 2015–June 2015 | 110 | 413 | 128 | 23 | 128 | ~ |
| Class 45 | July 2015–Dec. 2015 | 110 | 388 | 101 | 12 | 101 | 3 |

#### Physical Fitness

|  | Curl-Ups | | Push-Ups | | 1-Mile Run | |
|---|---|---|---|---|---|---|
|  | Initial | Final | Initial | Final | Initial | Final |
| Class 44 | 34.4 | 57.0 | 21.3 | 46.3 | 09:51 | 08:03 |
| Class 45 | 38.4 | 45.2 | 23.3 | 36.5 | 09:59 | 08:56 |

#### Responsible Citizenship

|  | Voting | | Selective Service | |
|---|---|---|---|---|
|  | Eligible | Registered | Eligible | Registered |
| Class 44 | 13 | 13 | 37 | 32 |
| Class 45 | 9 | 9 | 25 | 25 |

#### Service to Community

|  | Hours of Service | Dollar Value/Hr | Total Value |
|---|---|---|---|
| Class 44 | 8,911 | $21.50 | $191,581 |
| Class 45 | 6,824 | $21.50 | $146,716 |

#### Postresidential Performance Status

|  | Graduated | Contacted | Placed | Education | Employment | Military | Other |
|---|---|---|---|---|---|---|---|
| Class 44 |  |  |  |  |  |  |  |
| Month 1 | 128 | 128 | 96 | 81 | 8 | 0 | 7 |
| Month 6 | 128 | 128 | 105 | 86 | 18 | 1 | 0 |
| Month 12 | 128 | 127 | 93 | 68 | 23 | 2 | 0 |
| Class 45 |  |  |  |  |  |  |  |
| Month 1 | 101 | 102 | 89 | 81 | 7 | 1 | 0 |
| Month 6 | 101 | 102 | 82 | 64 | 15 | 1 | 2 |

~ Does not award

**Table C.28**
**Oregon Profile**

| OREGON YOUTH CHALLENGE PROGRAM, ESTABLISHED 1999 |
|---|

| Graduates since inception: 3,861 | Program type: High School Diploma, GED or HiSET, Credit Recovery |
|---|---|

### Staffing

| | Instructional | Cadre | Administrative | Other | Total |
|---|---|---|---|---|---|
| Full-time | 5 | 27 | 21 | 0 | 53 |
| Part-time | 0 | 0 | 0 | 0 | 0 |

### Funding

| | Federal Funding | State Funding |
|---|---|---|
| Classes 44 and 45 | $3,770,000 | $1,256,667 |

### Residential Performance

| | Dates | Target | Applied | Graduated | Received GED/ HiSET | Received HS Credits | Received HS Diploma |
|---|---|---|---|---|---|---|---|
| Class 44 | Jan. 2015–June 2015 | 120 | 198 | 125 | 6 | 125 | 28 |
| Class 45 | July 2015–Dec. 2015 | 120 | 247 | 134 | 4 | 134 | 22 |

### Physical Fitness

| | Curl-Ups | | Push-Ups | | 1-Mile Run | |
|---|---|---|---|---|---|---|
| | Initial | Final | Initial | Final | Initial | Final |
| Class 44 | 37.6 | 55.4 | 18.4 | 29.2 | 12:08 | 12:06 |
| Class 45 | 37.5 | 52.9 | 16.3 | 30.3 | 12:09 | 12:07 |

### Responsible Citizenship

| | Voting | | Selective Service | |
|---|---|---|---|---|
| | Eligible | Registered | Eligible | Registered |
| Class 44 | 38 | 38 | 61 | 61 |
| Class 45 | 42 | 42 | 83 | 83 |

### Service to Community

| | Hours of Service | Dollar Value/Hr | Total Value |
|---|---|---|---|
| Class 44 | 11,043 | $22.75 | $251,223 |
| Class 45 | 12,773 | $22.75 | $290,574 |

### Postresidential Performance Status

| | Graduated | Contacted | Placed | Education | Employment | Military | Other |
|---|---|---|---|---|---|---|---|
| Class 44 | | | | | | | |
| Month 1 | 125 | 109 | 88 | 42 | 53 | 0 | 8 |
| Month 6 | 125 | 109 | 97 | 73 | 30 | 2 | 2 |
| Month 12 | 125 | 117 | 105 | 67 | 30 | 4 | 2 |
| Class 45 | | | | | | | |
| Month 1 | 134 | 130 | 120 | 106 | 26 | 0 | 1 |
| Month 6 | 134 | 112 | 96 | 81 | 28 | 3 | 1 |

**Table C.29**
**Puerto Rico Profile**

| PUERTO RICO YOUTH CHALLENGE ACADEMY, ESTABLISHED 1999 |
|---|

| Graduates since inception: 4,871 | Program type: High School Diploma |
|---|---|

**Staffing**

|  | Instructional | Cadre | Administrative | Other | Total |
|---|---|---|---|---|---|
| Full-time | 11 | 49 | 38 | 9 | 107 |
| Part-time | 0 | 0 | 0 | 0 | 0 |

**Funding**

|  | Federal Funding | State Funding |
|---|---|---|
| Classes 44 and 45 | $3,500,000 | $1,166,667 |

**Residential Performance**

|  | Dates | Target | Applied | Graduated | Received GED/ HiSET | Received HS Credits | Received HS Diploma |
|---|---|---|---|---|---|---|---|
| Class 44 | Apr. 2015–Sept. 2015 | 200 | 334 | 225 | ~ | 225 | 225 |
| Class 45 | Oct. 2015–Mar. 2016 | 200 | 307 | 225 | ~ | 225 | 225 |

**Physical Fitness**

|  | Curl-Ups | | Push-Ups | | 1-Mile Run | |
|---|---|---|---|---|---|---|
|  | Initial | Final | Initial | Final | Initial | Final |
| Class 44 | 31.8 | 41.3 | 23.7 | 37.8 | 09:12 | 07:33 |
| Class 45 | 35.7 | 44.3 | 27.5 | 39.0 | 08:29 | 07:28 |

**Responsible Citizenship**

|  | Voting | | Selective Service | |
|---|---|---|---|---|
|  | Eligible | Registered | Eligible | Registered |
| Class 44 | 49 | 49 | 44 | 44 |
| Class 45 | 67 | 67 | 53 | 53 |

**Service to Community**

|  | Hours of Service | Dollar Value/Hr | Total Value |
|---|---|---|---|
| Class 44 | 13,360 | $11.39 | $152,170 |
| Class 45 | 11,504 | $11.39 | $131,031 |

**Postresidential Performance Status**

|  | Graduated | Contacted | Placed | Education | Employment | Military | Other |
|---|---|---|---|---|---|---|---|
| Class 44 |  |  |  |  |  |  |  |
| Month 1 | 225 | 225 | 95 | 35 | 31 | 0 | 29 |
| Month 6 | 225 | 224 | 188 | 140 | 36 | 2 | 10 |
| Month 12 | 225 | 224 | 193 | 129 | 57 | 2 | 5 |
| Class 45 |  |  |  |  |  |  |  |
| Month 1 | 225 | 225 | 103 | 41 | 35 | 0 | 27 |
| Month 6 | 225 | 225 | 201 | 165 | 30 | 1 | 5 |

~ Does not award

**Table C.30**
**South Carolina Profile**

| SOUTH CAROLINA YOUTH CHALLENGE ACADEMY, ESTABLISHED 1998 |
|---|

Graduates since inception: 3,129        Program type: GED

### Staffing

|  | Instructional | Cadre | Administrative | Other | Total |
|---|---|---|---|---|---|
| Full-time | 6 | 27 | 26 | 0 | 59 |
| Part-time | 1 | 1 | 1 | 0 | 3 |

### Funding

|  | Federal Funding | State Funding |
|---|---|---|
| Classes 44 and 45 | $2,800,000 | $1,000,000 |

### Residential Performance

|  | Dates | Target | Applied | Graduated | Received GED/ HiSET | Received HS Credits | Received HS Diploma |
|---|---|---|---|---|---|---|---|
| Class 44 | Jan. 2015–June 2015 | 100 | 129 | 94 | 21 | 2 | 1 |
| Class 45 | July 2015–Dec. 2015 | 100 | 262 | 103 | 47 | 1 | * |

### Physical Fitness

|  | Curl-Ups | | Push-Ups | | 1-Mile Run | |
|---|---|---|---|---|---|---|
|  | Initial | Final | Initial | Final | Initial | Final |
| Class 44 | 38.6 | 44.5 | * | * | * | 08:02 |
| Class 45 | 35.0 | 49.9 | 9.2 | 11.8 | 09:40 | 07:59 |

### Responsible Citizenship

|  | Voting | | Selective Service | |
|---|---|---|---|---|
|  | Eligible | Registered | Eligible | Registered |
| Class 44 | 32 | 31 | 27 | 26 |
| Class 45 | 24 | 24 | 23 | 23 |

### Service to Community

|  | Hours of Service | Dollar Value/Hr | Total Value |
|---|---|---|---|
| Class 44 | 5,246 | $21.14 | $110,900 |
| Class 45 | 5,586 | $21.14 | $118,088 |

### Postresidential Performance Status

|  | Graduated | Contacted | Placed | Education | Employment | Military | Other |
|---|---|---|---|---|---|---|---|
| Class 44 |  |  |  |  |  |  |  |
| Month 1 | 94 | 96 | 63 | 6 | 23 | 0 | 34 |
| Month 6 | 94 | 96 | 79 | 43 | 18 | 0 | 18 |
| Month 12 | 94 | 96 | 79 | 43 | 18 | 0 | 18 |
| Class 45 |  |  |  |  |  |  |  |
| Month 1 | 103 | 103 | 76 | 56 | 15 | 0 | 5 |
| Month 6 | 103 | 103 | 80 | 44 | 28 | 0 | 8 |

* Did not report

**Table C.31**
**Texas, East Profile**

### TEXAS CHALLENGE ACADEMY—EAST, ESTABLISHED 2014

Graduates since inception: 51

Program type: Credit Recovery, High
School Diploma, GED or HiSET

**Staffing**

|  | Instructional | Cadre | Administrative | Other | Total |
|---|---|---|---|---|---|
| Full-time | 10 | 26 | 16 | 6 | 58 |
| Part-time | 1 | 0 | 0 | 0 | 1 |

**Funding**

|  | Federal Funding | State Funding |
|---|---|---|
| Classes 44 and 45 | $2,560,000 | $853,333 |

**Residential Performance**

|  | Dates | Target | Applied | Graduated | Received GED/ HiSET | Received HS Credits | Received HS Diploma |
|---|---|---|---|---|---|---|---|
| Class 44 | ^ | ^ | ^ | ^ | ^ | ^ | ^ |
| Class 45 | July 2015–Dec. 2015 | 100 | 156 | 51 | 24 | 50 | 12 |

**Physical Fitness**

|  | Curl-Ups | | Push-Ups | | 1-Mile Run | |
|---|---|---|---|---|---|---|
|  | Initial | Final | Initial | Final | Initial | Final |
| Class 44 | ^ | ^ | ^ | ^ | ^ | ^ |
| Class 45 | * | * | * | * | * | * |

**Responsible Citizenship**

|  | Voting | | Selective Service | |
|---|---|---|---|---|
|  | Eligible | Registered | Eligible | Registered |
| Class 44 | ^ | ^ | ^ | ^ |
| Class 45 | 15 | 15 | 15 | 15 |

**Service to Community**

|  | Hours of Service | Dollar Value/Hr | Total Value |
|---|---|---|---|
| Class 44 | ^ | ^ | ^ |
| Class 45 | 1,811 | $25.11 | $45,474 |

**Postresidential Performance Status**

|  | Graduated | Contacted | Placed | Education | Employment | Military | Other |
|---|---|---|---|---|---|---|---|
| Class 44 |  |  |  |  |  |  |  |
| Month 1 | ^ | ^ | ^ | ^ | ^ | ^ | ^ |
| Month 6 | ^ | ^ | ^ | ^ | ^ | ^ | ^ |
| Month 12 | ^ | ^ | ^ | ^ | ^ | ^ | ^ |
| Class 45 |  |  |  |  |  |  |  |
| Month 1 | 51 | 51 | 32 | 27 | 13 | 2 | 2 |
| Month 6 | 51 | 51 | 40 | 26 | 21 | 0 | 2 |

* Did not report

^ Newly operational

**Table C.32**
**Texas, West Profile**

| TEXAS CHALLENGE ACADEMY—WEST, ESTABLISHED 1999 |
|---|

Graduates since inception: 2,921

Program type: High School Diploma, GED or HiSET, Credit Recovery

### Staffing

|  | Instructional | Cadre | Administrative | Other | Total |
|---|---|---|---|---|---|
| Full-time | 9 | 24 | 24 | 0 | 57 |
| Part-time | 0 | 0 | 0 | 0 | 0 |

### Funding

|  | Federal Funding | State Funding |
|---|---|---|
| Classes 44 and 45 | $2,400,000 | $800,000 |

### Residential Performance

|  | Dates | Target | Applied | Graduated | Received GED/ HiSET | Received HS Credits | Received HS Diploma |
|---|---|---|---|---|---|---|---|
| Class 44 | Jan. 2015–June 2015 | 100 | 213 | 93 | 43 | 93 | 9 |
| Class 45 | July 2015–Dec. 2015 | 100 | 164 | 74 | 42 | 74 | 4 |

### Physical Fitness

|  | Curl-Ups | | Push-Ups | | 1-Mile Run | |
|---|---|---|---|---|---|---|
|  | Initial | Final | Initial | Final | Initial | Final |
| Class 44 | 34.5 | 47.0 | 24.3 | 51.8 | 12:20 | 10:38 |
| Class 45 | 34.9 | 47.5 | * | * | 09:56 | 07:23 |

### Responsible Citizenship

|  | Voting | | Selective Service | |
|---|---|---|---|---|
|  | Eligible | Registered | Eligible | Registered |
| Class 44 | 23 | 23 | 34 | 32 |
| Class 45 | 29 | 29 | 41 | 41 |

### Service to Community

|  | Hours of Service | Dollar Value/Hr | Total Value |
|---|---|---|---|
| Class 44 | 4,093 | $25.11 | $102,763 |
| Class 45 | 3,150 | $25.11 | $79,084 |

### Postresidential Performance Status

|  | Graduated | Contacted | Placed | Education | Employment | Military | Other |
|---|---|---|---|---|---|---|---|
| Class 44 |  |  |  |  |  |  |  |
| Month 1 | 93 | 89 | 15 | 5 | 10 | 0 | 0 |
| Month 6 | 93 | 84 | 49 | 43 | 5 | 1 | 0 |
| Month 12 | 93 | 85 | 65 | 38 | 23 | 4 | 0 |
| Class 45 |  |  |  |  |  |  |  |
| Month 1 | 74 | 70 | 23 | 18 | 5 | 0 | 0 |
| Month 6 | 74 | 70 | 42 | 24 | 16 | 2 | 0 |

* Did not report

**Table C.33**
**Virginia Profile**

### VIRGINIA COMMONWEALTH CHALLENGE YOUTH ACADEMY, ESTABLISHED 1994

Graduates since inception: 4,387

Program type: Credit Recovery, GED or HiSET

#### Staffing

|  | Instructional | Cadre | Administrative | Other | Total |
|---|---|---|---|---|---|
| Full-time | 9 | 40 | 21 | 0 | 70 |
| Part-time | 0 | 0 | 2 | 0 | 2 |

#### Funding

|  | Federal Funding | State Funding |
|---|---|---|
| Classes 44 and 45 | $3,375,000 | $1,548,470 |

#### Residential Performance

|  | Dates | Target | Applied | Graduated | Received GED/ HiSET | Received HS Credits | Received HS Diploma |
|---|---|---|---|---|---|---|---|
| Class 44 | Oct. 2014–Feb. 2015 | 135 | 163 | 66 | 10 | 11 | ~ |
| Class 45 | Mar. 2015–Aug. 2015 | 135 | 183 | 91 | 28 | 25 | ~ |

#### Physical Fitness

|  | Curl-Ups | | Push-Ups | | 1-Mile Run | |
|---|---|---|---|---|---|---|
|  | Initial | Final | Initial | Final | Initial | Final |
| Class 44 | 44.5 | * | 33.4 | * | 09:08 | * |
| Class 45 | 36.9 | 50.7 | 25.7 | 53.1 | 09:09 | 07:37 |

#### Responsible Citizenship

|  | Voting | | Selective Service | |
|---|---|---|---|---|
|  | Eligible | Registered | Eligible | Registered |
| Class 44 | 21 | 21 | 30 | 28 |
| Class 45 | 22 | 22 | 47 | 46 |

#### Service to Community

|  | Hours of Service | Dollar Value/Hr | Total Value |
|---|---|---|---|
| Class 44 | 2,716 | $26.09 | $70,860 |
| Class 45 | 7,162 | $26.09 | $186,857 |

#### Postresidential Performance Status

|  | Graduated | Contacted | Placed | Education | Employment | Military | Other |
|---|---|---|---|---|---|---|---|
| Class 44 |  |  |  |  |  |  |  |
| Month 1 | 66 | 66 | 52 | 34 | 25 | 0 | 1 |
| Month 6 | 66 | 66 | 52 | 32 | 36 | 0 | 1 |
| Month 12 | 66 | 66 | 42 | 23 | 28 | 1 | 0 |
| Class 45 |  |  |  |  |  |  |  |
| Month 1 | 91 | 91 | 69 | 53 | 30 | 1 | 1 |
| Month 6 | 91 | 91 | 63 | 45 | 34 | 3 | 2 |

* Did not report

~ Does not award

**Table C.34**
**Washington Profile**

| WASHINGTON YOUTH ACADEMY, ESTABLISHED 2009 |
|---|

| Graduates since inception: 1,615 | Program type: Credit Recovery |
|---|---|

**Staffing**

|  | Instructional | Cadre | Administrative | Other | Total |
|---|---|---|---|---|---|
| Full-time | 6 | 36 | 34 | 0 | 76 |
| Part-time | 3 | 4 | 1 | 0 | 8 |

**Funding**

|  | Federal Funding | State Funding |
|---|---|---|
| Classes 44 and 45 | $3,600,000 | $1,200,000 |

**Residential Performance**

|  | Dates | Target | Applied | Graduated | Received GED/ HiSET | Received HS Credits | Received HS Diploma |
|---|---|---|---|---|---|---|---|
| Class 44 | Jan. 2015–June 2015 | 125 | 225 | 140 | ~ | 140 | ~ |
| Class 45 | July 2015–Dec. 2015 | 125 | 264 | 152 | ~ | 152 | ~ |

**Physical Fitness**

|  | Curl-Ups | | Push-Ups | | 1-Mile Run | |
|---|---|---|---|---|---|---|
|  | Initial | Final | Initial | Final | Initial | Final |
| Class 44 | 46.4 | 73.2 | 17.2 | 44.9 | 10:45 | 07:39 |
| Class 45 | 52.7 | 66.7 | 17.2 | 39.1 | 09:30 | 08:38 |

**Responsible Citizenship**

|  | Voting | | Selective Service | |
|---|---|---|---|---|
|  | Eligible | Registered | Eligible | Registered |
| Class 44 | 60 | 60 | 52 | 52 |
| Class 45 | 48 | 48 | 51 | 51 |

**Service to Community**

|  | Hours of Service | Dollar Value/Hr | Total Value |
|---|---|---|---|
| Class 44 | 7,810 | $28.99 | $226,397 |
| Class 45 | 7,292 | $28.99 | $211,389 |

**Postresidential Performance Status**

|  | Graduated | Contacted | Placed | Education | Employment | Military | Other |
|---|---|---|---|---|---|---|---|
| Class 44 |  |  |  |  |  |  |  |
| Month 1 | 140 | 140 | 102 | 79 | 23 | 0 | 0 |
| Month 6 | 140 | 140 | 122 | 112 | 10 | 0 | 0 |
| Month 12 | 140 | 140 | 105 | 87 | 15 | 2 | 1 |
| Class 45 |  |  |  |  |  |  |  |
| Month 1 | 152 | 152 | 137 | 132 | 4 | 0 | 1 |
| Month 6 | 152 | 152 | 138 | 133 | 4 | 0 | 1 |

~ Does not award

**Table C.35**
**Wisconsin Profile**

### WISCONSIN CHALLENGE ACADEMY, ESTABLISHED 1998

Graduates since inception: 3,200

Program type: GED or HiSET, High School Diploma, Credit Recovery, Other

#### Staffing

|  | Instructional | Cadre | Administrative | Other | Total |
|---|---|---|---|---|---|
| Full-time | 5 | 23 | 18 | 0 | 46 |
| Part-time | 0 | 10 | 1 | 3 | 14 |

#### Funding

|  | Federal Funding | State Funding |
|---|---|---|
| Classes 44 and 45 | $3,569,594 | $1,189,915 |

#### Residential Performance

|  | Dates | Target | Applied | Graduated | Received GED/ HiSET | Received HS Credits | Received HS Diploma |
|---|---|---|---|---|---|---|---|
| Class 44 | Jan. 2015–June 2015 | 100 | 262 | 103 | 53 | * | * |
| Class 45 | July 2015–Dec. 2015 | 100 | 267 | 100 | 48 | * | * |

#### Physical Fitness

|  | Curl-Ups | | Push-Ups | | 1-Mile Run | |
|---|---|---|---|---|---|---|
|  | Initial | Final | Initial | Final | Initial | Final |
| Class 44 | 18.4 | 46.0 | 13.2 | 26.0 | 10:09 | 07:45 |
| Class 45 | 18.2 | 42.9 | 12.1 | 21.0 | 09:42 | 08:08 |

#### Responsible Citizenship

|  | Voting | | Selective Service | |
|---|---|---|---|---|
|  | Eligible | Registered | Eligible | Registered |
| Class 44 | 33 | 33 | 57 | 54 |
| Class 45 | 31 | 31 | 55 | 55 |

#### Service to Community

|  | Hours of Service | Dollar Value/Hr | Total Value |
|---|---|---|---|
| Class 44 | 7,241 | $22.48 | $162,783 |
| Class 45 | 7,539 | $22.48 | $169,471 |

#### Postresidential Performance Status

|  | Graduated | Contacted | Placed | Education | Employment | Military | Other |
|---|---|---|---|---|---|---|---|
| Class 44 |  |  |  |  |  |  |  |
| Month 1 | 103 | 103 | 57 | 9 | 46 | 0 | 5 |
| Month 6 | 103 | 103 | 74 | 27 | 59 | 1 | 4 |
| Month 12 | 103 | 103 | 73 | 14 | 63 | 5 | 2 |
| Class 45 |  |  |  |  |  |  |  |
| Month 1 | 100 | 100 | 66 | 42 | 34 | 0 | 4 |
| Month 6 | 100 | 100 | 80 | 29 | 62 | 1 | 5 |

* Did not report

**Table C.36**
**West Virginia Profile**

| MOUNTAINEER CHALLENGE ACADEMY, ESTABLISHED 1993 |
|---|

Graduates since inception: 3,431

Program type: High School Diploma, Credit Recovery

**Staffing**

|  | Instructional | Cadre | Administrative | Other | Total |
|---|---|---|---|---|---|
| Full-time | 6 | 28 | 14 | 10 | 58 |
| Part-time | 1 | 0 | 0 | 0 | 1 |

**Funding**

|  | Federal Funding | State Funding |
|---|---|---|
| Classes 44 and 45 | $3,383,100 | $1,125,000 |

**Residential Performance**

|  | Dates | Target | Applied | Graduated | Received GED/ HiSET | Received HS Credits | Received HS Diploma |
|---|---|---|---|---|---|---|---|
| Class 44 | Jan. 2015–June 2015 | 100 | 305 | 138 | 116 | 137 | 116 |
| Class 45 | July 2015–Dec. 2015 | 125 | 394 | 152 | 145 | 151 | 145 |

**Physical Fitness**

|  | Curl-Ups | | Push-Ups | | 1-Mile Run | |
|---|---|---|---|---|---|---|
|  | Initial | Final | Initial | Final | Initial | Final |
| Class 44 | 33.0 | 53.7 | * | * | 10:04 | 07:08 |
| Class 45 | 33.1 | 55.3 | * | * | 09:49 | 07:38 |

**Responsible Citizenship**

|  | Voting | | Selective Service | |
|---|---|---|---|---|
|  | Eligible | Registered | Eligible | Registered |
| Class 44 | 32 | 32 | 31 | 31 |
| Class 45 | 34 | 34 | 28 | 28 |

**Service to Community**

|  | Hours of Service | Dollar Value/Hr | Total Value |
|---|---|---|---|
| Class 44 | 6,477 | $20.47 | $132,574 |
| Class 45 | 10,126 | $20.47 | $207,269 |

**Postresidential Performance Status**

|  | Graduated | Contacted | Placed | Education | Employment | Military | Other |
|---|---|---|---|---|---|---|---|
| Class 44 |  |  |  |  |  |  |  |
| Month 1 | 138 | 137 | 14 | 0 | 14 | 0 | 0 |
| Month 6 | 138 | 137 | 62 | 13 | 46 | 3 | 0 |
| Month 12 | 138 | 137 | 74 | 6 | 55 | 11 | 2 |
| Class 45 |  |  |  |  |  |  |  |
| Month 1 | 152 | 152 | 23 | 3 | 19 | 0 | 1 |
| Month 6 | 152 | 152 | 87 | 4 | 65 | 17 | 1 |

* Did not report

**Table C.37**
**Wyoming Profile**

## WYOMING COWBOY CHALLENGE ACADEMY, ESTABLISHED 2005

| Graduates since inception: 771 | Program type: GED or HiSET, Credit Recovery |
|---|---|

### Staffing

| | Instructional | Cadre | Administrative | Other | Total |
|---|---|---|---|---|---|
| Full-time | 6 | 25 | 21 | 0 | 52 |
| Part-time | 0 | 0 | 0 | 0 | 0 |

### Funding

| | Federal Funding | State Funding |
|---|---|---|
| Classes 44 and 45 | $1,500,000 | $2,008,337 |

### Residential Performance

| | Dates | Target | Applied | Graduated | Received GED/HiSET | Received HS Credits | Received HS Diploma |
|---|---|---|---|---|---|---|---|
| Class 44 | Jan. 2015–June 2015 | 80 | 125 | 64 | 46 | 8 | 4 |
| Class 45 | July 2015–Dec. 2015 | 80 | 123 | 58 | 32 | 5 | 2 |

### Physical Fitness

| | Curl-Ups | | Push-Ups | | 1-Mile Run | |
|---|---|---|---|---|---|---|
| | Initial | Final | Initial | Final | Initial | Final |
| Class 44 | 28.9 | 38.5 | 30.8 | 49.0 | 09:05 | 07:58 |
| Class 45 | 28.0 | 34.8 | 26.1 | 33.1 | 09:15 | 08:24 |

### Responsible Citizenship

| | Voting | | Selective Service | |
|---|---|---|---|---|
| | Eligible | Registered | Eligible | Registered |
| Class 44 | 9 | 9 | 16 | 10 |
| Class 45 | 9 | 9 | 18 | 9 |

### Service to Community

| | Hours of Service | Dollar Value/Hr | Total Value |
|---|---|---|---|
| Class 44 | 2,980 | $23.13 | $68,927 |
| Class 45 | 2,880 | $23.13 | $66,614 |

### Postresidential Performance Status

| | Graduated | Contacted | Placed | Education | Employment | Military | Other |
|---|---|---|---|---|---|---|---|
| Class 44 | | | | | | | |
| Month 1 | 64 | 64 | 32 | 7 | 25 | 0 | 0 |
| Month 6 | 64 | 64 | 38 | 13 | 25 | 0 | 0 |
| Month 12 | 64 | 64 | 38 | 8 | 26 | 2 | 2 |
| Class 45 | | | | | | | |
| Month 1 | 58 | 58 | 35 | 16 | 17 | 0 | 2 |
| Month 6 | 58 | 58 | 40 | 19 | 17 | 4 | 0 |

# Abbreviations

| | |
|---|---|
| ABE/ASE | Adult Basic Education/Adult Secondary Education |
| ACT | Standardized College Entrance Exam (formerly known as American College Testing) |
| AFQT | Armed Forces Qualifying Test |
| ChalleNGe | National Guard Youth ChalleNGe Program |
| CPI-U | Consumer Price Index for All Urban Consumers |
| ESEA | Elementary and Secondary Education Act |
| GAIN | General Assessment of Instructional Needs |
| GED | General Educational Development |
| HiSET | High School Equivalency Test |
| HS | high school |
| HSLS | High School Longitudinal Study |
| M&E | monitoring and evaluation |
| MAPT | Massachusetts Adult Proficiency Test |
| NCLB | No Child Left Behind |
| NRS | National Reporting Service for Adult Education |
| NSC | National Student Clearinghouse |
| OCTAE | Office of Career, Technical and Adult Education |
| SAT | Scholastic Aptitude Test |
| SLDS | Statewide Longitudinal Data System |
| TABE | Tests of Adult Basic Education |
| TOC | theory of change |
| UI | unemployment insurance |
| USDOE | U.S. Department of Education |

# References

"Adult Education—CalEdFacts," *California Department of Education*, May 15, 2017. As of June 23, 2017:
http://www.cde.ca.gov/sp/ae/po/cefadulted.asp

"American Community Survey (ACS)," *United States Census Bureau*, March 17, 2017. As of June 23, 2017:
https://www.census.gov/programs-surveys/acs/news/data-releases.2014.html

Amin, Vikesh, Carlos A. Flores, Alfonso Flores-Lagunes, and Daniel Parisian, "The Effect of Degree Attainment on Arrests: Evidence from a Randomized Social Experiment," *Economics of Education Review*, Vol. 54, 2016, pp. 259–273.

Anderson, Andrea, *The Community Builder's Approach to Theory of Change: A Practical Guide to Theory and Development*, New York: Aspen Institute Roundtable on Community Change, 2005.

Bruce, Mary, and John Bridgeland, *The Mentoring Effect: Young People's Perspectives on the Outcomes and Availability of Mentoring*, Washington, D.C.: Civic Enterprises with Peter D. Hart Research Associates, 2014.

Card, David, "The Casual Effect of Education on Earnings," in Orley C. Ashenfelter and David Card, eds., *Handbook of Labor Economics*, Vol. 3, Part A, 1999, pp. 1801–1863.

CASAS—*See* Comprehensive Adult Student Assessment System.

City Colleges of Chicago, *Testing and Placement Guidelines*, Chicago, 2015. Retrieved from:
http://www.ccc.edu/menu/ . . . /Testing%20and%20Placement%202015%20Policy.docx

Code of Federal Regulations, Title 34, Sections 461–463, Adult Education, 1998.

Comprehensive Adult Student Assessment System, "Study of the CASAS Relationship to GED 2002: Research Brief," San Diego, Calif., June 2003. As of January 13, 2017:
https://www.casas.org/docs/default-source/research/download-what-is-the-relationship-between-casas
-assessment-and-ged-2002-.pdf?sfvrsn=5?Status=Master

———, "Study of the CASAS Relationship to GED 2014: Research Brief," San Diego, Calif., March 2016. As of January 13, 2017:
https://www.casas.org/docs/default-source/research/study-of-the-casas-relationship-to-ged-2014.pdf

"Consumer Price Index," *Bureau of Labor Statistics*, n.d. As of January 13, 2017:
http://www.bls.gov/cpi/

"Core Components," *National Youth Guard ChalleNGe*, n.d. As of January 13, 2017:
https://www.jointservicessupport.org/NGYCP/core-components/EeqO-z8MgUK9NFRHRW8flQ

Dalton, Ben, Steven J. Ingles, Laura Fritch, and Elise Christopher, *High School Longitudinal Study of 2009 (HSLS:09)*, Washington, D.C.: U.S. Department of Education, NCES 2015-037rev, 2016. As of December 13, 2016:
http://nces.ed.gov/pubs2015/2015037rev.pdf

Dee, Thomas, "A Teacher Like Me: Does Race, Ethnicity, or Gender Matter?" *American Economic Review*, Vol. 95, No. 2, 2005, pp. 158–165.

Division of Adult Education and Literacy Office of Vocational and Adult Education, *Implementation Guidelines: Measures and Methods for the National Reporting System for Adult Education*, Washington, D.C.: U.S. Department of Education, 2015. As of January 13, 2017:
http://www.nrsweb.org/docs/implementationguidelines.pdf

Eaton, Danice K., Laura Kann, Steve Kinchen, Shari Shanklin, James Ross, Joseph Hawkins, William A. Harris, Richard Lowry, Tim McManus, David Chyen, Connie Lim, Lisa Whittle, Nancy Brener, and Howell Wechsler, *Youth Risk Behavior Surveillance—United States 2009*, Washington, D.C.: Centers for Disease Control and Prevention, Vol. 59 (SS05), June 4, 2010, pp. 1–142. As of November 8, 2016: http://www.cdc.gov/MMWR/Preview/MMWRhtml/ss5905a1.htm

"Economic News Release," *Bureau of Labor Statistics*, January 6, 2017. As of January 13, 2017: http://www.bls.gov/news.release/empsit.t01.htm

"Every Student Succeeds Act (ESSA)," *U.S. Department of Education*, n.d. As of June 23, 2017: https://www.ed.gov/essa?src=ft

Federal Register, Vol. 80, Section 48304, Tests Determined to Be Suitable for Use in the National Reporting System for Adult Education, 2015.

Freeman, Jennifer, and Brandi Simonsen, "Examining the Impact of Policy and Practice Interventions on High School Dropout and School Completion Rates: A Systematic Review of the Literature," *Review of Educational Research*, Vol. 85, No. 2, 2015, pp. 205–248.

Gonzalez, Gabriela C., Laura L. Miller, and Thomas E. Trail, *The Military Spouse Education and Career Opportunities Program*, Santa Monica, Calif.: RAND Corporation, RR-1013-OSD, 2016. As of March 27, 2017: http://www.rand.org/pubs/research_reports/RR1013.html

Heckman, James, "Policies to Foster Human Capital," *Research in Economics*, Vol. 54, 2000, pp. 3–56.

Henrichson, Christian, and Ruth Delaney, *The Price of Prisons: What Incarceration Costs Taxpayers*, New York: Vera Institute of Justice, July 2012. As of January 13, 2017: http://archive.vera.org/sites/default/files/resources/downloads/price-of-prisons-updated-version-021914.pdf

Hossain, Farhana, and Dan Bloom, *Toward a Better Future: Evidence on Improving Employment Outcomes for Disadvantaged Youth in the United States*, New York: MDRC, 2015.

Ingels, Steven J., Daniel J. Pratt, Deborah R. Herget, Michael Bryan, Laura Burns Fritch, Randolph Ottem, James E. Rogers, and David Wilson, *High School Longitudinal Study of 2009 (HSLS:09) 2013 Update and High School Transcripts Data File Documentation*, Washington, D.C.: National Center for Education Statistics, 2013. As of December 13, 2016: https://nces.ed.gov/surveys/hsls09/hsls09_data.asp

Institute of Education Sciences, "SLDS Topical Webinar Summary: Linking K12 Education Data to Workforce," 2014. As of January 13, 2017: https://nces.ed.gov/programs/slds/pdf/Linking_K12_Education_Data_to_Workforce_August2014.pdf

Jacob, Brian A., and Jesse Rothstein, "The Measurement of Student Ability in Modern Assessment Systems," *Journal of Economic Perspectives*, Vol. 30, No. 3, 2016, pp. 85–108.

Jacobson, Neil S., and Paula Truax, "Clinical Significance: A Statistical Approach to Defining Meaningful Change in Psychotherapy Research," *Journal of Consulting and Clinical Psychology*, Vol. 59, 1991, pp. 12–19.

Knowlton, Lisa Wyatt, and Cynthia C. Phillips, *The Logic Model Guidebook: Better Strategies for Great Results*, Thousand Oaks, Calif.: Sage, 2009.

Lindholm-Leary, Kathryn, and Gary Hargett, *Evaluator's Toolkit for Dual Language Programs*, Sacramento, Calif.: California Department of Education, December 2006. As of January 13, 2017: http://www.lindholm-leary.com/toolkit2/toolkit2index.htm

Malone, Lauren, and Jennifer Atkin, *Cognitive and Noncognitive Improvements among ChalleNGe Cadets: A Survey of Seven Sites*, Arlington, Va.: Center for Naval Analyses, DRM 2016-U-013175, 2016. As of December 6, 2016: https://www.cna.org/CNA_files/PDF/DRM-2016-U-013175-Final.pdf

Massachusetts Department of Elementary and Secondary Education, Adult and Community Learning Services, *Assessment Policies for Using the TABE Forms 9/10 Test*. Malden, Mass., 2015. As of June 23, 2017: http://www.doe.mass.edu/acls/assessment/TABEpolicy.pdf

McConnell, Sheena, and Steven Glazerman, *National Job Corps Study: The Benefits and Costs of Job Corps*, Washington, D.C.: Mathematica Policy Research, 2001.

Millenky, Megan, Dan Bloom, Sara Muller-Ravett, and Joseph Broadus, *Staying on Course: Three-Year Results of the National Guard Youth ChalleNGe Evaluation*, New York: MDRC, 2011.

National Center for Education Statistics, "Total and Current Expenditures per Pupil in Public Elementary and Secondary Schools: Selected Years, 1919–20 Through 2013–14," *Digest of Education Statistics*, Table 236.55, July 2016. As of January 13, 2017:
https://nces.ed.gov/programs/digest/d16/tables/dt16_236.55.asp?current=yeshttps://nces.ed.gov/programs/digest/d16/tables/dt16_236.55.asp?current=yes

National Guard Youth ChalleNGe, home page, n.d. As of January 13, 2017:
https://www.jointservicessupport.org/NGYCP/

———, *2015 Performance and Accountability Highlights*, Arlington, Va.: National Guard Bureau, December 2015. As of November 6, 2016:
http://www.people.mil/Portals/56/Documents/Reports/2015%20NGYCP%20Annual%20Report%20Final.pdf?ver=2016-09-09-153221-517

"National Guard Youth ChalleNGe," *National Mentoring Resource Center*, n.d. As of January 13, 2017:
http://www.nationalmentoringresourcecenter.org/index.php/insight-display/58-national-guard-youth-challenge.html

National Reporting Service for Adult Education, "NRS Test Benchmarks for Educational Functioning Levels," Washington, D.C., 2015.

———, "NRS Tips: Sampling for the Follow-Up Surveys," Washington, D.C., n.d. As of January 13, 2017:
http://www.nrsweb.org/docs/NRSTips_Sampling_FINAL_508.pdf

New York State Education Department, *New York State Assessment Policy*, Albany, N.Y., 2015. As of June 23, 2017:
http://www.acces.nysed.gov/common/acces/files/aepp/NYSAssessmentPolicyRevisedApril2015.docx

NRS—*See* National Reporting Service for Adult Education.

Olsen, M., *Guide to Administering TABE (Tests of Adult Basic Education): A Handbook for Teachers and Test Administrators*. Little Rock, Ark.: Arkansas Department of Career Education, 2009. As of January 13, 2017:
http://ace.arkansas.gov/adulteducation/documents/tabehandbook2009.pdf

Perez-Arce, Francisco, Louay Constant, David S. Loughran, and Lynn A. Karoly, *A Cost-Benefit Analysis of the National Guard Youth ChalleNGe Program*, Santa Monica, Calif.: RAND Corporation, TR-1193-NGYF, 2012. As of January 13, 2017:
http://www.rand.org/pubs/technical_reports/TR1193.html

Price, Hugh, "Foundations, Innovation and Social Change: A Quixotic Journey Turned Case Study," working paper presented during practitioner residency, Rockefeller Foundation Bellagio Center, 2010. As of December 2016:
http://cspcs.sanford.duke.edu/sites/default/files/Foundations%20Innovation%20and%20Social%20Change.pdf

Rennie-Hill, Leslie, Jenni Villano, Michelle Feist, Nettie Legters, Jean Thomases, and Patrice Williams, *Bringing Students Back to the Center: A Resource Guide for Implementing and Enhancing Re-Engagement Centers for Out-of-School Youth*. Washington, D.C.: U.S. Department of Education, 2014.

Rhodes, Jean E., Renee Spencer, Thomas E. Ketter, Belle Liang, and Gil Noam, "A Model for the Influence of Mentoring Relationships on Youth Development," *Journal of Community Psychology*, Vol. 34, No. 6, 2006, pp. 691–707.

Schwartz, Sarah E. O., Jean E. Rhodes, Renee Spencer, and Jean B. Grossman, "Youth Initiated Mentoring: Investigating a New Approach to Working with Vulnerable Adolescents," *American Journal of Community Psychology*, Vol. 52, 2013, pp. 155–169.

Shakman, Karen, and Sheila M. Rodriguez, *Logic Models for Program Design, Implementation, and Evaluation: Workshop Toolkit*, Washington, D.C.: U.S. Department of Education, 2015. As of December 12, 2016:
http://files.eric.ed.gov/fulltext/ED556231.pdf

"TABE Tests of Adult Basic Education," *Data Recognition Corporation*, n.d. As of January 13, 2017:
http://www.datarecognitioncorp.com/assessment-solutions/pages/tabe.aspx

Tierney, Joseph P., Jean B. Grossman, and Nancy L. Resch, *Making a Difference: An Impact Study of Big Brothers Big Sisters*, Philadelphia: Public/Private Ventures, 2000.

Topel, Robert H., and Michael P. Ward, "Job Mobility and the Careers of Young Men," NBER Working Paper No. 2649, 1988.

U.S. Code 32, Chapter Five, Section 509, National Guard Youth ChalleNGe Program of Opportunities for Civilian Youth, Part K, "Report."

U.S. Department of Education, Office of Career, Technical, and Adult Education, *Adult Education and Family Literacy Act of 1998: Annual Report to Congress, Program Year 2011–12*, Washington, D.C., 2015.

U.S. Department of Education, Office of Career, Technical, and Adult Education, Division of Adult Education and Literacy, *Implementation Guidelines: Measures and Methods for the National Reporting System for Adult Education*. Washington, D.C., 2016. As of June 23, 2017:
http://www.nrsweb.org/docs/NRS_Implementation_Guidelines_February2016.pdf

USDOE—*See* U.S. Department of Education

Wenger, Jennie, Cathleen McHugh, and Lynda Houck, *Attrition Rates and Performance of ChalleNGe Participants over Time*, Arlington, Va.: Center for Naval Analyses, CRM D0013758.A2/Final, 2006.

Wenger, Jennie, Cathleen McHugh, Seema Sayla, and Robert Shuford, *Variations in Participants and Policies Across ChalleNGe Programs*, Arlington, Va.: Center for Naval Analyses, CRM D0016643.A2/Final, 2008.

West Virginia Department of Education, *Correlation Between Various Placement Instruments for Reading, Language/Writing, Mathematics, Elementary Algebra*, Charleston, W.Va., n.d. As of June 23, 2017:
https://wvde.state.wv.us/abe/documents/CorrelationBetweenVariousPlacementInstruments.pdf

W. K. Kellogg Foundation, *Using Logic Models to Bring Together Planning, Evaluation, and Action: Logic Model Development Guide*, Battle Creek, Mich., 2006.